D0251062

A Healthy, Wealthy, Sustainable World

A Healthy, Wealthy, Sustainable World

John Emsley

RSC Publishing

ISBN: 978-1-84755-862-6

A catalogue record for this book is available from the British Library

Published by The Royal Society of Chemistry,
Thomas Graham House, Science Park, Milton Road,
Cambridge CB4 0WF, UK

Registered Charity Number 207890

For further information see our web site at www.rsc.org

Foreword

I am delighted to introduce prospective readers to *A Healthy, Wealthy, Sustainable World.* Currently, there is a considerable appetite amongst students of all ages to improve their knowledge of science in general, and chemistry in particular. Thus, people's natural curiosity has been stimulated by the considerable media coverage of environmental concerns, especially debates on global warming and the challenge of maintaining economic growth whilst sustaining life on Earth, given its finite resources and the impact of human activity on ecosystems and the environment. Vested interests, prejudice and ignorance often lead to political, commercial or personal choices that adversely affect the quality of life on this planet. This book provides a wealth of scientific information in an interesting and very accessible form. Thus, it should dispel much of the ignorance and, hopefully, remove some of the prejudice that compromise constructive debate and rational decision making. Such matters have considerable significance for all of us, especially on behalf of future generations; if we do not find a solution to the problems that we face, a solution will find us, and other forms of life will flourish on Earth.

This book comprises seven chapters, each of which provides a fascinating and comprehensive account of the vital roles of chemistry in Food, Water, Health, Transport, Plastics, Cities and Sport. The text strikes a remarkably good balance between presenting information with clarity and authority whilst retaining the reader's interest, especially by highlighting the relevance of the information provided to every day life. In particular, I congratulate the author on providing an informed and objective account of *chemicals*. On the one hand he shows that life is a series of chemical events, describes the essential roles of chemicals in healthcare and nutrition, the provision of clean water, and in the production of objects for our material world. On the other, he details the toxic nature of many chemicals and how these have lead to injury, disease and death and caused significant environmental problems. This book is an excellent reference source and provides a wealth of information with references and footnotes in the text, together with a comprehensive glossary and many suggestions for further reading.

All readers, whatever the level and nature of their scientific expertise, will significantly improve their appreciation and knowledge of each of the topics covered. In summary, this is an excellent book that should be essential reading in all secondary schools and used as a reference for courses and projects in tertiary education. Furthermore, it is to the author's credit that the book is also ideal for reading on a train journey or in the armchair.

C. David Garner
University of Nottingham, UK

Preface

[A word in **bold** means there is more information in the Glossary.]

AS I SEE IT

For most people the issue of sustainability is generally presented in the form of finding alternative sources of energy, rarely in the form of sustainability regarding the other materials we produce from fossil reserves, although this a problem which chemists often talk about. Green chemistry will be essential if a green lifestyle is not to be a materially impoverished lifestyle, with the products of the chemical industry no longer available. Food, clean water, fuel, healing drugs and plastics might then be in short supply. The challenge of moving from fossil carbon to **biomass** carbon is not going to be easy but it is not impossible. What it will need is a new generation of bright young chemists to bring it about. This book is meant to encourage young people to engage with this science, but it is not written only for them. If you are someone who wishes to know more about the way that chemistry underpins so much of modern life then this is the book for you.

The ancient Greek philosophers reasoned that there were four elements and these were earth, air, fire and water. The ancient Chinese had a fifth: wood. These were not elements of course – in that respect they were wrong – but they are the major categories which define the world as we know it, namely land, atmosphere, energy, water and biomass, and these are the 'elements' that comprise the world's current problems.

Although many people don't realise it, and some perhaps don't even like to acknowledge it, our developed lifestyle depends to a great extent on chemistry and its associated industries. Two hundred years ago, life for our forebears was short, dirty, disease-ridden, painful and at times blighted with hunger. Today it is longer, cleaner, healthier, mostly

A Healthy, Wealthy, Sustainable World
By John Emsley

Published by the Royal Society of Chemistry, www.rsc.org

pain-free and with food in abundance, thanks in part to the science of chemistry. The object of this book is to explain the part chemistry plays in our developed-world lifestyle and to explore whether it is possible, at least in theory, to enjoy a similar standard of living relying only on sustainable resources.

Our basic needs of sufficient food, clean water, and healing drugs rely on the agrochemical, chemical and pharmaceutical industries, respectively. Other essential items of the good life, namely fuels and plastics, are more clearly dependent on fossil reserves and producing them sustainably is going to be a major challenge that chemists, biochemists and process engineers will have to solve this century. Meanwhile humans all over the world crowd into ever larger cities and chemistry can help make these more sustainable and more enjoyable.

Could a small country like the UK with its 60 million people produce all the food, fuel and materials it needs from the land it has available – and still leave space for wildlife? I believe it can and in so doing it would set the world an example of sustainable living. Of course if the human population of this planet were to stop expanding and even decline to say 5 billion – with the UK at 50 million – then this would help. Chemists have provided the means of bringing this about in the form of untearable condoms, birth control pills and even morning-after pills, but these are disapproved of by some. Even with these products available, the world's population is already 6.5 billion and heading towards 9 billion by the end of this century.

Brand image is vitally important to success, which means that chemistry has a problem. It has had a bad press for almost half a century. Even the word 'chemical' is now used as a negative term by many to signify something of which they disapprove. In 2008 the UK's Advertising Standards Agency (ASA) was asked to adjudicate on an advert for an 'organic' fertiliser which claimed it was '100% chemical-free'. The ASA said there was nothing wrong with the wording and that the claim could be made. Chemists were dumbstruck. Everything on this planet is 100% chemical when looked at from a chemist's viewpoint.

The propaganda of the green movement, organic farming and alternative medicine has always portrayed chemistry and chemicals in a poor light. Things got so bad in the 1990s that cancer, hyperactive children, low sperm counts in modern men, the sudden death of North Sea seals, and even the fall in the numbers of songbirds and bees were blamed on 'chemicals'. (All later were shown to be wrong.) So successful has the anti-chemical message been that some people now think there is a distinction to be made between *natural* chemicals, which are by definition safe, and *man-made* chemicals, which are dangerous. This opinion may be nonsense but it tarnishes the image of chemists. In this book I will try in a small way to re-brand chemistry. If we are to live in a sustainable world by the end of this century, what the world needs are *more* chemists and their industries, not fewer.

Toxicologists are the scientists who truly understand the effects of chemicals on the human body and their opinions have now been canvassed regarding the claims made by environmental groups about the dangers of 'chemicals'. To a chemist like myself it came as no surprise to find that 96% of the members of the US Society of Toxicology believed that Greenpeace's claims are grossly exaggerated, while 75% of them said the media gives too much prominence to the findings of individual researchers funded by environmental groups. In the case of phthalates, which have been particularly targeted by environmentalists, 89% of toxicologists said they did not regard phthalates as a high risk at all.

Some believe that the Enlightenment, the age of science and reason which started in Europe 350 years ago, is giving way to the Endarkenment in which irrational beliefs verging on superstition hold sway and nowhere more so than in the area of health and cures for illness. Alternative practitioners are everywhere offering unscientific treatments such as homeopathy, crystal healing, acupuncture, herbal medicine, aromatherapy, reflexology, naturopathy, nutritional therapy and spiritual healing. Of course some of them appear to produce results, just as placebos do. These alternatives may be little more than harmless diversions in a developed society but they can be highly detrimental for people in developing countries, as happens when herbal remedies are touted as being effective antimalarial drugs, when homeopathic treatments are advocated for TB, and when diet and vitamins are promoted as ways to protect and treat AIDS.

Roger Bacon, who lived in the 1200s, was a Franciscan friar and philosopher and he was the first to advocate the scientific method. He saw how easy it was for people to be fooled into believing what is not true. He listed four traps into which we could fall: (1) by accepting the opinions of a faulty authority, (2) by relying on established customs, (3) by following the vagaries of popular prejudice, and (4) by being unaware of our own limitations. But even if his approach encourages us to question what we are told, we still need to know whom to trust and what to believe. Science based on *repeated* experimentation is the only sure bedrock for accepting something as true. One-off experiments and the amateur epidemiology undertaken by those working for the alternative culture prove nothing at all.

Just as there is a placebo effect in which as many as a third of people in double-blind tests appear to benefit from a non-active agent, so there is the nocebo effect in which a large number of people appear to be adversely affected by what they perceive to be a dangerous substance, when in fact it is perfectly harmless. In the case of food additives, those known to be perfectly safe are entitled to carry an E number, but this has not stopped some groups labelling them as 'chemicals' and saying that they were a threat to health. Their faulty authority has resulted in lots of people now regarding E numbers as warnings rather than reassurances.

Every decade of the last century produced remarkable advances in the chemical and pharmaceutical industries. In the early 1900s there was the production of ammonia for fertilisers from nitrogen and hydrogen, on which a quarter of the world's population now depends for their food supply. In the 1930 and 1940s came a host of new polymers, such as polythene, Teflon, nylon and polyester, which are now part of everyday life. The 1950s and 1960s were the golden years of new drugs which saved millions of lives and made life longer and more bearable for millions of others. In the 1970s and 1980s came the materials which are now an integral part of modern technology, such as microchips and low-energy light bulbs. Then finally in the 1990s came chemistry's answers to environmental problems, such as water-based household paints, eco-friendly detergents and targeted pesticides. But as the last century bowed out, the chemical industries were no longer regarded favourably despite all the benefits they had brought.

If the 20th century was the one in which chemistry helped mould the modern world, what will its role be in the 21st century? Will it even have a role? Surely we already know everything there is to know about plastics and paints, containers and carpets, shower gels and skin creams, drugs and diets? To a certain extent this is true, and there are probably only minor adjustments that can be made to perfecting them, although of course there is a need to produce them sustainably. This will mean finding better catalysts, cleaner solvents, new technologies and novel processes. Oil refineries and chemical plants that currently occupy sites as large as small towns may one day stand on sites no larger than a football stadium – and they will produce no waste. That is the basis of the new green chemistry.

Chemists this century will need to come up with new methodologies to support the expanding biotechnology, pharmaceutical and micro-electronics industries. We may see crops and biomass producing all kinds of raw materials as resources for industry but these will still need to be transformed into useable products. If cancer is ever to be conquered it will surely be through the combined efforts of medics, biochemists and chemists. But even if they come up with the answer, how will the new drug be manufactured and how will it be delivered into the body? From production to patient, the chemical-based industries will have a role to play. Perhaps the most exciting area of chemistry to emerge in the future will be new materials for other industries. Imagine a material that could convert sunlight into electricity with an efficiency of 50%, or an underground cable which could transmit electricity over large distances with no loss of power, say from solar energy farms in southern deserts to cold northern cities.

If the chemical industry of this century is finally to be based on sustainable resources then what are the basic raw materials which it must have? We will need to produce ethylene in vast quantities for various polymers, and that can come from ethanol which is already produced on a large scale as bioethanol. We will also need propylene for polymers and that can be made from glycerol which is a by-product of plant oils. Another major raw material is benzene which is needed for nylon and polystyrene, and that could in theory be extracted from the lignin of wood. There is no limit to what chemists can produce given a source of that all-important element, carbon, which is the basis of all biomass. What might be in short supply are brilliant young chemists to make it all possible.

The concept of sustainable development came from the UN's World Commission on Environment and Development (aka the Brundtland Commission) which was set up in 1983 and published its report in 1986. This defined sustainable development as "meeting the needs of the present without compromising the ability of future generations to meet their own needs". The message is finally getting through, helped by people like Paul Anastas and John Warner who in 1998 wrote *Green Chemistry: Theory & Practice* which set down certain principles on which a sustainable chemical industry could be based, such as waste prevention, atom economy,[1] safer solvents, energy efficiency, pollution prevention, *etc.* These guidelines are now being followed by the major companies.

Something happened around 2005, which has been seen as the tipping point, when people suddenly realised that the new century would have to be different in its use of **fossil fuels** if we were not to ruin the planet. While chemists had been talking about this for many years, the event which brought it to the attention of many was *Chemistry for a Sustainable Future,* a workshop organised by the US National Science

[1] Atom economy was the concept popularised by US chemist Barry Trost, and this says that for any chemical reaction producing a commercial product the aim should be atom in atom out, in other words incorporating all materials used in a process into the final product with no by-products or waste.

Foundation and held in May 2006. This recognised that chemistry was the science to lead to a developed yet sustainable planet.

The World Business Council for Sustainable Development is an association of 200 companies from 35 different countries and includes Air Products & Chemicals, Bayer, China Petrochemical, Dow Chemical, DuPont, Mitsubishi Chemical, Novartis, Novo Nordisk, Rohm & Haas, Syngenta, as well as companies that market their products to consumers such as Procter & Gamble, S.C. Johnson & Son and Unilever. Some green organisations are now working with the chemical industry.

The chemicals industry of the 20th century transformed the world. The chemical industry of the 21st century has to repeat the miracle if our grandchildren are to enjoy a future free of hunger, dirt, disease and material poverty. I will try and show that it is possible; otherwise we will stay on our current journey of going to hell in a green handcart. Maybe that's the destination we deserve. Sustainable chemical heaven may be unattainable – but at least we could give it a try.

THINGS TO NOTE

Throughout the book there are boxes entitled 'Common Sense', in which I show that popular beliefs about various chemical topics are little more than urban myths or 'factoids', as the American novelist Norman Mailer called them. He defined them as "something that everyone knows is true, except it aint".

The word 'organic' has two diametrically opposed meanings. To the general public it means food produced in ways which do not involve products of the agrochemical industry. To the chemist it means molecules based on carbon, and indeed the vast majority of the molecules which chemists have made are organic. In this book I need to use organic in both ways and to avoid confusion I will use 'organic' in inverted commas when referring to the non-chemical term.

Throughout the text some words are given in bold and this indicates more information to be found in the Glossary. It does not include molecular formulae which are readily accessible on Wikipedia, but it will include alternative names from chemicals and some linear formulae.

If you live in the UK, you might be disconcerted to discover I have spelled sulfur as sulfur, and similarly with sulfite and sulfuric acid. These are not misprints but are their preferred spellings as laid down by the International Union of Pure and Applied Chemistry (IUPAC). If you are a chemist you might be disconcerted to discover that I often use the common name of many chemicals – plastics instead of polymers, for instance – and this has been done deliberately because I also want this book to be read by people who are not chemists, but who are nevertheless educated and willing to learn about an area of life that is important if their grandchildren are to live in a sustainable world. It's their language I'm using because it's their children who will create that world.

John Emsley, DSc, FRSC

Contents

A Healthy, Wealthy, Sustainable World
By John Emsley

Published by the Royal Society of Chemistry, www.rsc.org

Acknowledgements

A Healthy, Wealthy, Sustainable World is about food, water, fuel, medicine, plastics, cities and sport, and the part which chemists will play in ensuring their future. Its span is so wide that I consulted experts in these various fields to check what I had written. The following people kindly came to my assistance. To all of them I owe a great debt of gratitude.

Stuart Dunbar of Syngenta sampled Chapter 1 on food, especially the sections on fertilisers and pesticides and he also read the section on the use of water in agriculture in Chapter 2. Richard Ratcliffe, Executive Director of the Food Additives and Ingredients Association, also digested Chapter 1. Richard Burt, who worked in the Food Standards Agency and was past President of the Institute of Food Science and Technology, also partook of Chapter 1 and made some valuable comments on its ingredients.

John Lindeman BSc is a Fellow of the Water Management Society and a Member of the Chartered Institute of Water and Environmental Management, and he waded through Chapter 2, as did Alistair Steel and Véronique Garny of Euro Chlor, the Brussels-based organisation which represents the chlor-alkali industry. David Wickens, Quality & Environment Manager of Severn Trent Water, and Stuart Bell, Dish-washing Manager of Reckitt Benckiser, also helped clean up Chapter 2.

Alistair Crawford MD gave Chapter 3 a check-up as did Dr Michael Utidjian, one-time Corporate Medical Director of the former company American Cyanamid.

James Clark, Professor of Chemistry and Director of the Green Chemistry Centre at York University, drove through Chapter 4 on biofuels and suggested a few better routes. Dr Michael Utidjian gave it a test ride.

Both Philip Law, Public & Industrial Affairs Director, and Sarah Plant, Industrial Issues Director, of the British Plastics Federation inspected the plastics of Chapter 5 for flaws.

Steve James, Managing Director, and Philip Hall, Technical Manager, of Techtron, an industrial maintenance chemical company based in Aldridge, West Midland, toured Chapter 6 to check that all was in order. Tim Jones, Professor of Chemistry at Warwick University, viewed the section on solar sources and flat screens.

Tony Kingsbury of ClearEurope refereed all of Chapter 7 and spotted a few off-side comments. Tennis enthusiast, Emeritus Professor Ray Richards, and former Manchester University sports champion, umpired the chapter and declared it game, set and match. Véronique Garny of Euro Chlor checked that the swimming pool section conformed to regulations. Phillip Glynn-Davies, Crash Worthiness Manager, and colleagues Graham Hobbs and Matthew Hillam of Millbrook Proving Ground in Bedfordshire put Chapter 7 through its paces and especially the section on Formula 1 racing. When it came to horse doping then Angela Beaver, a scientist within the Sport Science section of the Horseracing Forensic Laboratory (HFL), checked the course.

Finally a word of thanks to my wife Joan and to my good friends Steve and Rose Ley who were prevailed upon to read the whole of the finished text.

TRADE NAMES

CHAPTER 1

Food and Chemistry

[A word in **bold** means there is more information in the Glossary.]

The way to produce more food on less land and to do it sustainably requires chemistry. Here we look at four topics relating to food where chemistry is vital: fertilisers, pesticides, preservatives and additives. In addition, three topics are discussed where an understanding of chemistry explains how some components of food, other than the nutrients, can also affect us: natural toxins, neutraceuticals and food fraud.

During the time it takes you to read this sentence, the world's population will have increased by 20. By the time you have read this chapter, enough babies will have been born to populate a small town. In fact, the world's population will continue to grow for most of this century and may even reach 9 billion. They will all have to be fed.

According to the UN, the world population in 1950 was 2.5 billion living off 1.3 billion hectares of arable land. In 2000 there were more than twice as many people but, thanks to more productive agriculture, there was only a ten per cent increase in the area of farmed land. By 2030 there will be three times as many people and possibly no increase in the area of land for food production, although by then there may be twice as much land devoted to biofuel crops (which is the topic of Chapter 4). If no one is to go hungry we must farm scientifically, and that must include genetically modified crops. Even those will need fertilisers and pesticides, and the food from them will need preservatives so that it doesn't go bad before we can eat it.

A Healthy, Wealthy, Sustainable World
By John Emsley

Published by the Royal Society of Chemistry, www.rsc.org

Many have been told to suspect all things 'chemical' because this means it is unnatural. This use of the word 'chemical' betrays a scientific ignorance. I hope that by the time you have read this chapter you will appreciate that what chemists are mostly doing is reinforcing what Nature already does.

Maybe one day there will be GM (genetically modified) crops which make their own nitrogen fertiliser; produce more of their edible components; provide the range of nutrients which humans need; and resist all forms of pest and microbe attacks. However, until that happy day we will have to rely on chemistry to do many of these things, and that chemistry must be sustainable. We need fertilisers whose manufacture requires much less energy; agents to boost plant growth to maximise output; pesticides to protect crops; and preservatives to ensure the edible parts remain safe to eat after storage.

Common Sense 1: 'Organic' farming is the sustainable way to produce food

Correct, in a world of few people living in small groups. Wrong, in the world of towns and cities. 'Organic' farming is not without its good points in that it recycles manure and compost, but it has its limitations which far outweigh these benefits when it comes to feeding 9 billion people.

The advocates of 'organic' food say that crops should only be fertilised with manure and compost, rather than artificial fertilisers, and that chemical pesticides should be avoided. They even claim that 'organic' food is healthier food and 'organic' farming is sustainable. In fact, it is neither.

'Organic' farming cannot feed the people of today's world because of its lower yields of crops. 'Organic' farming needs at least a third more land than conventional farming. It is unsustainable as a way of feeding people in towns and cities because it cannot replace the soil nutrients that crops remove and without which the fertility of the land slowly decreases, as happened in ages past in several parts of the world. Even the use of animal manure and plant compost to replenish the soil is only at the expense of nutrients being taken from the other land which produces them.

For Europe to feed itself entirely on organically grown food would mean ploughing up an extra 25 million hectares of land, equivalent to an area somewhat larger than Wales.

We eat and drink to provide our body with the energy we need for warmth and movement, and the chemicals it needs to repair and renew damaged and worn out tissue. The nutrients it needs are carbohydrates, proteins, fats, vitamins, minerals and, of course, water. Most of what we eat comes from plants either directly or indirectly *via* animals, and plants also need nutrients. The first topic in this chapter is all about providing these in the form of fertilisers, and the most important one is nitrogen.

1.1 FERTILISERS

*Without the **ammonium nitrate** produced by the chemical industry the world would starve.*

Local famines, even national famines, have threatened humans with starvation throughout history. Could we one day face a global famine? Worldwide hunger was envisaged and warned against in the past by Thomas Malthus (1766–1834). Malthus predicted that there must eventually be a catastrophe for the human race because we reproduce 'geometrically', in other words we were multiplying, whereas food production could only increase 'arithmetically', in other words by the simple addition of more land devoted to farming.[1] There must come a time when there were just too many mouths to feed and no more land to farm and there would be mass starvation. This didn't happen, however, because in the 1800s the plains of North America began to produce food in abundance much of which was exported.

Worries about overpopulation surfaced again in the 1960s as numbers continued to increase. In 1968, Paul Ehrlich's book *The Population Explosion* predicted a future crisis with hundreds of millions of people dying of starvation. Another group to warn of an impending food crisis was an influential NGO (non-governmental organisation) called The Club of Rome. In 1972 their report, entitled *The Limits to Growth*, forecast that that there would be severe food shortages by the year 2000. That crisis was averted by the application of even better agrochemicals and better strains of the major food crops. Today the majority of people in the world get enough to eat, but many get barely enough to survive and stand little chance of making much of their lives. Eventually all will need to be fed by sustainable agriculture which will have to rely on a sustainable agrochemical industry.

[1] Geometric growth can be seen as doubling, as in the sequence 1,2,4,8,16,32 . . . , while arithmetic growth is simply an increase by the addition of the same number each time, as in the sequence 1,2,3,4,5,6 . . . The gap between these two sequences becomes large very quickly.

Figure 1.1 Thanks to chemistry there's so much more food to harvest these days.

A plant needs nutrients to grow and the elements it needs most are carbon, nitrogen, potassium and phosphorus. The first of these is not a problem because plants take their carbon from the carbon dioxide in the air, and there is enough of that as we know. The other three have to come from the soil and the one which most limits plant growth is nitrogen. Why is this? The reason is that all living things need **amino acids** which have nitrogen as part of their molecular make-up. These acids link together to form proteins and enzymes, and the result is that the average person contains 2 kg of nitrogen.

Plants take in nitrogen as either ammonia or nitrate, and these are in the soil and come from the debris of dead plants, animal excreta, and soil fauna and flora. A little nitrate even arrives with rainwater, which dissolves the nitrogen oxides produced in thunderstorms and from vehicle exhausts. There is plenty of nitrogen on Earth because it constitutes around 80% of the atmosphere, totalling an incredible 4 trillion tonnes, but this is mainly inaccessible to plants. However, some plants can draw on atmospheric nitrogen, such as legumes like peas and beans, thanks to the rhizobia bacteria which are present in the root nodules of these plants. The rhizobia take **carbohydrate** from the plant in return for soluble nitrogen salts which they produce. Most plants rely on the microbes in the soil breaking down plant and animal residues for their ammonium and nitrate, however, and this source of nitrogen is limited and can become exhausted. The key to boosting crop yields is to add ammonium nitrate from another source.

Early attempts by chemists to react nitrogen and hydrogen failed to produce any ammonia (NH_3), no matter how strongly they were heated together. What the reaction needed was a catalyst and that's what the German chemist Fritz Haber discovered,[2] and chemical engineer Carl Bosch turned it into a commercial process which is carried out at around 450 °C and under high pressure. On 3rd July 1909, the first successful Haber–Bosch chemical plant started up at Oppau, Germany. Today, there are Haber–Bosch plants around the world producing 150 million tonnes of ammonia a year, the vast majority of which goes into making fertilisers. Two billion people worldwide are now reliant on this source of agrochemical nitrogen for their food supply. Industry produces almost as much 'fertiliser' nitrogen as the biomass fixes naturally.

Nitrate is made from ammonia by reacting the gas with oxygen and water. There is also a process for making nitrate from the oxygen and nitrogen in the atmosphere, known as the Birkeland–Eyde process, which was patented by two Norwegians in the early 1900s and operated successfully for many years. Using a powerful electric discharge to mimic the effect of lightning, nitrogen and oxygen react together to form nitric oxide (NO) which then reacts with more oxygen and water to form nitrate, as nitric acid (HNO_3).

Combine ammonia and nitric acid and you have ammonium nitrate. Spread or spray a dilute solution of this on the land when a growing crop needs it and you can double, triple and even quadruple the yield. Opponents of the agrochemical industry call such fertiliser 'artificial' (*i.e.* meaning unnatural), but of course it matters not to the roots of a plant whether the ammonia or nitrate comes from rotting matter or from the chemical industry.

At the present time the production of ammonium nitrate is not sustainable because the energy to make it comes from fossil fuels. Sometime in the future we will either have to genetically modify plants so that, like legumes, they can support rhizomes on their roots, or we must devise a way of manufacturing ammonium nitrate using only renewable energy. It may be possible one day to do it on a local scale with the Birkeland–Eyde process using energy derived from a waterwheel or windmill. If there were such a unit in every farming community, the environmental impact would be minimal and crop yields would remain large and sustainable.

[2]The catalyst is iron activated by potassium hydroxide. Haber's first catalysts were either osmium or uranium which worked spectacularly well, but were too expensive.

In the 1980s there were scare stories based on **epidemiology** about nitrate from fertilisers getting into the water supply and causing stomach cancer and the scary sounding **blue-baby syndrome**, although the last case of this in the UK was 60 years ago. Happily 11 other studies showed no link, and 7 even showed a negative correlation, in other words more nitrate meant *fewer* cases of cancer. Indeed, supposedly 'healthy' foods like lettuce, spinach, beetroot, celery and potatoes have naturally high levels of nitrate, and eating lots of those has never been linked to cancer. In 1985 it was discovered that the human body even generates its own nitrate at around 70 mg per day, similar to that coming from food and water. Cells release nitrate in response to infections and even to strenuous physical exercise such as running and cycling. Nitrate protects the human body against disease pathogens.

No benefit comes totally without cost and so it is with nitrogen fertilisers. Overuse leaves this in the soil, there to be acted upon by microbes which release some of the nitrogen as nitrous oxide (N_2O), a powerful greenhouse gas. Indeed the world is using much too much ammonium nitrate to fertilise crops – some estimates suggest that we could make do with less than half of that which we now use if there were better management strategies, such as tailoring application to crop needs. This has been undertaken in the Netherlands where a 40% reduction of nitrogen fertilisers has had little effect on crop yields.

And what of the other macro nutrients that crops need? These are phosphate and potassium. Are they sustainable?

Human sewage and animal manure represent an important source of phosphate. Both can be processed to recover phosphate especially that from large intensive livestock farms or chicken sheds which can be turned into struvite (ammonium magnesium phosphate). This can be used as a slow-release fertiliser. Sewage is less easy to process because it is greatly diluted with water. Nevertheless, there is a growing need to find a valid use for the phosphate that has to be removed from it. Phosphate can be precipitated as an insoluble form, either as iron phosphate or aluminium phosphate, but these are unsuitable as fertiliser because the phosphate is chemically too tightly bound for plant roots to extract it.

Biological phosphorus reclamation was first suggested in 1955 when it was observed that aerated sewage sludge absorbed phosphate. The first biological process was known as the Phostrip process, and the sludge that settled from such treatment was suitable for use as an agricultural fertiliser. The bacteria with an appetite for phosphate are *Acinetobacter*, *Aeromonas* and *Pseudomonas*, There are now several commercial systems of biological phosphorus removal in use. As yet, the recycling of reclaimed phosphate for other uses is in its infancy, but research has shown that it is feasible.

Potassium can also be sustainable. In the 1800s the demand for potassium was met from forest clearances in the US. Trees were burned and their ash stirred with water to extract the water-soluble potassium salts. The solution so obtained was heated to dryness in large pots to yield potassium carbonate, hence its common name pot-ash. Most potassium for fertiliser is currently obtained by mining the minerals sylvite (potassium chloride) and carnallite (potassium and magnesium chlorides) of which there are deposits in excess of 10 billion tonnes, which means that sustainability is never likely to be an issue. Even so, potassium for fertilisers could be derived in part from the ash left after burning biomass.

1.2 PESTICIDES

Modern pesticides are our best weapons in the fight to protect the food we grow.

Let's start with a simple fact: half the fruit and vegetables grown by farmers would be unfit for market if pesticides were not used. The yields of 'organic' crops are typically 30% less than normal, and in the case of 'organic' potatoes is 40% less, which is part of the reason they are so expensive. Erich-Christian Oerke of the University of Bonn in Germany has demonstrated that without the products of the agrochemical industry, crop yields would fall to about half and crops like cotton would fall by as much as 80%. Protection of our food supply is needed at all stages, starting with the soil itself which is a hostile environment of pests and microbes. Protection is needed for seeds, for growing plants and for the harvested crop. All are under attack.

There are those who would ban pesticides because they think the traces of pesticides in their food represent a risk to their health. These people are prepared to pay premium prices for 'organic' food in the mistaken belief that this is produced without the need for chemical pesticides – it isn't, as we shall see. Pesticide residues are blamed for causing a variety of conditions ranging from lethargy to liver cancer, including several illnesses whose cause is as yet unknown. When it comes to modern pesticides these claims of damaging human health don't stand up to close inspection.

Those working for agrochemical companies make many new molecules which have potential as crop protection agents, but very few ever reach the farmer. Indeed, it has been calculated that only one compound in 100 000 will end up in the market place, and then only after years of testing to make sure it works and is safe. Consequently it costs around £150 million to launch a new pesticide. The hurdles it must pass include effectiveness; testing in various formulations and application routes; and biological testing against a variety of insects, weeds or diseases on various crops under a variety of conditions. It is also necessary to know how the pesticide actually works and what its breakdown products are in the soil and in plants. Only when all this data has been amassed, and approved by numerous agencies, will a licence be issued. Only large companies have the resources to invest in this area.[3]

Pesticides must not harm those who come into contact with them, must only destroy the target pests, and must be short-lived and break down into harmless substances. Pesticides in the past often failed in one of more of these aspects. What chemists also do is produce better targeted pesticides, so less is needed. Ideally, a modern pesticide will not bio-accumulate as did some of the early ones, like DDT.[4] In assessing whether a pesticide poses a risk, the guideline is the No-Observable-Adverse-Effect (NOAE) level. When this has been determined, then the amount that could possibly be consumed is set at 100 times less. Crop protection now requires much less insecticide than was once used. For example, the traditional herbicide 2,4-D has to be applied at a rate of 1 kg per hectare, whereas the best modern herbicides require only around 0.01 kg (10 g) per hectare. Some, such as glyphosate, will affect only plants because it blocks an enzyme which only plants have. The enzyme is used by plants to make plant protein, whereas animals make protein in a different way.

Plants produce their own pesticides to protect themselves against microbes and insects – and even animals. In fact most crop protection chemicals are produced naturally by plants. The world-famous molecular biologist Dr Bruce Ames, who devised the test for assessing the effects of chemicals on living things, pointed out that we ingest 10 000 times as many natural crop protection products than we do those made by the chemical industry. Some are remarkably effective, such as the products produced in the leaves of the neem tree of India. The leaves of this tree can even resist attack by a plague of locusts; their secret lies in the compound azadirachtin which actually deters an insect from further feeding once it has take a bite. In 2007 Steve Ley of Cambridge

[3] These are Syngenta, BASF, Dow, DuPont, Bayer CropScience, Sumitomo, Monsanto, Mitsui, Kumiai, Ishihara Sangyo Kaisha, Nihon Nohyakhu, Nissan and Nippon Soda.

[4] Although DDT was banned in many western countries for more than 30 years, it was approved again as an insecticide by the UN in 2006.

Common Sense 2: 'Organic' farmers don't use chemical pesticides

Wrong. They use a lot. Putting up a sign saying 'This is an Organic Farm' does not stop crop pests from entering and attacking whatever is growing there. Just like normal farmers, 'organic' farmers need to protect their crops and this involves the use of chemicals. For example, they use calcium polysulfide in apple orchards where it is sprayed at a rate of 13 kg per hectare per year, and their potatoes need it to be sprayed several times a season to counteract late blight fungus, while strawberry growers use as much as 40 kg per hectare if they are not to lose two thirds of their crop.

Other pesticides which 'organic' farmers use are **Bordeaux mixture** (copper sulfate plus calcium hydroxide), iron(III) orthophosphate and nicotine. This last one would never pass the health and safety tests demanded of modern pesticides because it is a deadly poison when ingested. Another 'organic' pesticide is the traditional derris dust, a broad-spectrum pesticide used against insects and extracted from the *Derris elliptica* plant which grows in SE Asia. It is highly effective because it contains the toxic chemical **rotenone**. However, it has been removed from the list of approved pesticides in the USA because it was shown to cause Alzheimer's type damage in the brains of rats, as well as being deadly to fish. Whether traces of this pesticide on 'organic' produce ever caused Alzheimer's in humans we will never know.

University Chemistry Department showed how azadirachtin could be produced chemically.

An example of a modern pesticide is indoxacarb, which was discovered in 1972 by Dutch researchers Rudolph Mulder and Kobus Wellinga, and kills only the target pest, in this case caterpillars. These little beasts are some of the most voracious crop feeders which attack cotton, soyabean, cabbage, peppers, tomatoes and alfalfa, which is a legume grown throughout the world mainly as cattle fodder. In 2000 DuPont launched indoxacarb. This chemical interferes with the caterpillar's ability to chew. It does not affect other creatures, such as beneficial insects, because the victim has to ingest the pesticide for it to kill. Indoxacarb itself is not the molecule which kills; the toxin which kills it is formed from indoxacarb by the caterpillar's own metabolic processes.

If potatoes are attacked by Colorado beetles they increase the amount of a defensive glycoalkaloid toxin in their tissue. However, this toxin is not always completely effective. The German chemical company BASF has launched Alverde. This new insecticide is based on metaflumizone and is effective against the Colorado beetle, which has developed resistance to some of the current crop of insecticides. This will be a niche insecticide which may also find use by vets and in homes, but will not be registered for general use on crops.

Why do insects become resistant to a particular herbicide or pesticide? In any population of plants and insects there will always be that rare individual which survives exposure to a toxic agent. For example, if the pesticide works by blocking a specific enzyme then it might be that this one individual has a slightly different arrangement of atoms around the enzyme's active site and this prevents the toxin gaining access. That being so then its descendents will also have that protection and within a few years a significant population of resistant insects will defeat the efforts of the farmer to eradicate them. However, the resistant gene has to be fit-for-purpose or it will fail to spread among the population. More worrying, is that the resistant gene not only blocks this particular pesticide but may block other pesticides as well.

The chemist's answer is to identify the cause of the resistance and then construct a new toxin that will defeat it. An alternative way to wage war against pests is to keep changing the pesticide and not always to rely on the same one. Another pesticide might well kill the insects carrying the mutant gene.

Chemical pesticides remain our main line of defence. For example, unsprayed fields of wheat can show as much as 80% wheat rust infection whereas sprayed crops have less than 2% infection. The potato crop is particularly vulnerable to attack, sometimes with terrible consequences. The late potato blight of 1846 in Ireland was caused by the fungus *Phytophthora infestans* and resulted in 1.5 million people dying of starvation. In a wet summer, fungus can still destroy around 15% of the crop. The traditional defence against fungal diseases of this kind was **Bordeaux mixture**. A much more effective agent is **dithane** (aka manganese zinc dithiocarbamate). Recently even this has been bettered with Revus (aka mandipropamid), which was discovered by chemists at Syngenta. Field trials were undertaken on potatoes infected with a virulent form of potato blight. Some plants were left untreated, some were sprayed with conventional fungicides, and some were sprayed with Revus. The field of potatoes was a desert of dead plants except for those sprayed with the new pesticide which were perfectly healthy.

Another powerful fungicide is based on material extracted from the fungus *Strobilurus tenacellus*. This fungicide fights its competitors with the natural chemical strobilium, but it is unsuitable as a man-made pesticide because it is sensitive to light, so chemists made more stable versions. These pesticides are needed to protect against soy rust, a major crop threat in countries like Brazil. Another new insecticide, spirotetramat, comes from Bayer and works by penetrating the plant to be protected. The insecticide is carried around in the sap so that it kills only sucking insects like aphids and whitefly, as well as those insects which attack its roots.

Farmers are now recommended to leave undisturbed margins around their fields where no pesticides are used in order to promote natural biodiversity and maybe even to harbour insects that would prey on the pest they are trying to eradicate. Moreover, farmers who use pesticides have now to be qualified to ensure they know how to dilute them and use them properly. (Pesticides for use on allotments and in home gardens are sold as ready made solutions.)

As well as chemical pesticides there are also biopesticides, whose sales now reach $1 billion. These may be predatory insects or microorganisms. The best known is the microbial pesticide *Bacillus thuringiensis* (aka Bt) which controls caterpillars and mosquito larvae. Bt is formulated as a powder and is applied to leaves where the insect larvae feed. It does not affect other organisms. Another biopesticide is *Bacillus subtilis* which has fungicidal properties.

A different approach to controlling insect pests is based on sex pheromones which lure insects into traps. These were introduced in the 1990s against the codling moth, and by 2004 more than 50 000 hectares of US orchards had them in place. Yet another approach is to develop semio-chemicals (aka signal chemicals), which plants generate when attacked to tell other plants to produce more of their natural defence molecules. Genetic modification might one day be used to introduce this ability into other plants.

A harmonised system for pesticide registration across Europe was introduced in 1991. Only those pesticides which do not cause human ill-heath, affect wildlife or pollute water supplies when used properly are permitted. At the time, there were 950 pesticides approved for use in Europe when directive 91/414/EEC came into force; now there are around 400 and this includes new pesticides introduced since then. In the year 2000 EU crop-chemical legislation was tightened up. More chemicals were withdrawn often because small manufacturers could not afford the cost of re-testing. For example, farmers used to be able to protect their bean crops by spraying with the herbicide *Flex* to control weeds, but this herbicide was withdrawn because it would have cost too much to have it re-registered. *Flex* is still used in the US, and is one of around 800 chemicals registered for use in crop protection around the world.

Even tighter EU regulations are coming and the number of approved pesticides will fall even further, maybe to less than 100. The new pesticide legislation is to be based on hazard rather than risk. In other words, if a chemical might conceivably cause an adverse reaction it would be banned whether or not it has ever affected anyone. What will remove a pesticide from the approved list is whether it fails tests for being persistent in the soil; whether it triggers neurotoxicity (nerve damage) although that can only be demonstrated on animals; whether it is an endocrine disrupter (meaning it affects an animal's hormonal system); and whether it kills bees. Somewhat inconsistently, the European Commission has approved fungicides used by 'organic' farmers which don't meet these requirements. The British Crop Protection Council argues that the data on which this decision was based would be insufficient to get approval for a product of the agrochemical industry. How odd.

1.3 FOOD ADDITIVES

Food additives include preservatives, colorants, emulsifiers, stabilisers, sweeteners, acidity-regulators, anti-caking agents, humectants (which prevent food from drying out), modified **starches** (which provide a smooth texture), raising agents (like baking powders) and flavouring agents. Here we will look more closely at preservatives and colorants, because these are the ones most often accused of being harmful.

1.3.1 Preservatives

Even more food would be wasted if it were not for protective molecules.

Our attitude to food additives can often lead people to adopt conflicting opinions. They will accept without question 'natural' ingredients in their food, without realising these may be dangerous, while at the same time opposing 'chemical additives' which are not a threat and which are there to protect them. Only when an additive has been subjected to rigorous testing and proved safe can food manufacturers use it. The object of adding a preservative to food is to prevent the multiplication of invading and potentially dangerous microbes. Were additives not used then there would be many more cases of food poisoning, which can occasionally be so serious as to kill.

Figure 1.2 Safe to eat? Yes, if they contain preservatives and so are not a breeding ground for germs.

In Europe, those food additives regarded as safe are given an E number. Sadly, this guarantee of safety has been wilfully misrepresented by certain pressure groups, whose disapproval has infected many of the general public who now avoid foods with E numbers listed as ingredients. Consequently, some companies have reverted to calling additives by their traditional names which is also permitted. Additives are often the most expensive ingredients in a recipe so they tend to be used in the smallest amounts legally permitted. How much of a preservative is needed depends on things like the composition of the food and its intended shelf-life.

Our ancestors had ways of preserving food, such as drying, smoking, pickling, baking and salting. Some of these techniques we still use. The industrial revolution brought other methods of food preservation, such as canning, pasteurising and refrigeration in the 1800s. The 1900s saw freeze-drying, vacuum packing and irradiation as other ways of preserving food, although the last of these is rarely used because of the public's irrational fear concerning radioactivity. Preservatives also provoked a similar response in some people.

Today there are thousands of processed foods in every supermarket and many rely on preservatives, such as sulfite, which is to be found in wines and dried fruits; nitrite, which is in cured meats; and benzoates, in soft drinks. The first two of these are additives of long standing – the ancient Romans used sulfites for preserving wine. Vintners burnt sulfur to form sulfur dioxide whose fumes were absorbed into the fermenting vats of grape juice thereby forming sulfite. Today sulfur dioxide is also used to preserve dried fruits. It can cause a reaction in some individuals who will experience a tightening of the chest and a difficulty in breathing. Were sulfite to be proposed as a way of preserving food today it would not pass the stringent tests required for its approval.

Here I would like to talk about more modern preservatives based on **benzoic acid** and on **parabenzoic acid**. Both occur naturally in fruits and other plants; benzoic acid is present in cranberries, plums, cloves and cinnamon, and parabenzoic acid is present in strawberries. However, in themselves these are not particularly useful chemicals because they may not dissolve in a product that needs their protection. Benzoic acid (E210) kills microbes but is only slightly soluble in water, which is why sodium benzoate (E211), which is much more soluble, is used. In solution this exists in equilibrium with benzoic acid, which is the active agent. Benzoic acid is rapidly absorbed into the body and rapidly metabolised in the liver to form hippuric acid, which is easily excreted *via* the urine. In 2000 a definitive report on benzoic acid and sodium benzoate was issued by the International Programme of Chemical Safety (IPCS), which is a joint venture of the UN, the World Health Organisation and the International Labour Organisation. The IPCS concluded that neither compound was carcinogenic, which is what some food activists had claimed, and indeed in health terms they had no observed adverse effect even when consumed in relatively large amounts.

The same goes for parabenzoic acid derivatives (aka parabens). They too are perfectly safe and yet they have been the target of a vicious and untruthful campaign linking them to breast cancer. These preservatives are esters of the acid, in other words the acidic hydrogen has been replaced by an organic group, such as a methyl, ethyl or propyl group. This makes them blend better with a wider range of foods. These esters have the E numbers: E218, E214 and E216, respectively. One of them, E218, occurs naturally in blueberries. There have been claims that parabens could be endocrine (*i.e.* hormone) disrupters and laboratory studies showed that this was in theory possible, but the effects being observed were a *billion* times weaker than those of normal hormones.

The late Maurice Hanssen's book *E for Additives*, which first appeared in 1984, says that benzoic acid causes gastric irritation and neurological disorders and that parabens may cause a numbing effect on the mouth. Both types are blamed for causing hyperactivity in children. His book did a lot to raise the hyperactivity of adults when it came to food additives, but it was not an unbiased account; witness Hanssen's warnings about E300, pointing out that 'large doses may cause diarrhoea and/or dental erosion'. This is highly misleading because E300 is actually vitamin C, which is, in fact, essential to health. If you want a proper scientific account of food additives then consult *Essential Guide to Food Additives* by Victoria Emerton and Eugenia Choi, the third edition of which appeared in 2008.

1.3.2 Colorants

These chemicals have been wrongly blamed for causing hyperactivity.

There may be psychological reasons why we prefer food that is colourful, which would explain why the food colorant industry globally is worth $1.2 billion per year. Many people now suspect 'chemical' colorants saying that they would prefer natural colorants, and blaming the former for causing their off-spring to behave badly. The food industry has responded to these wishes, but are the chemical colorants of long-standing really to blame for hyperactivity in young children?

The history of food colorants is rather shameful. When *The Lancet* published a paper in 1831 called 'Poisonous Confectionery' it listed some pretty obnoxious food colorants that were used, such as red lead (lead oxide, Pb_3O_4) and vermilion (mercury sulfide, HgS). The Analytical Sanitary Commission's report of 1851 listed 15 metal compounds that were also being used as colorants, including Scheele's green (copper arsenite, $CuAsO_3$), chrome yellow (lead chromate, $PbCrO_4$) and Prussian blue (iron hexacyanoferrate, $Fe_7(CN)_{18}$). Their report suggested safer alternatives and laws were passed which banned these kinds of food adulterants. Others were soon to take their place.

In the late 1800s, molecules were discovered which delivered a palette of intense colours the like of which had never been seen. They were ideal dyes for printing, paints and fabrics. These so-called aniline dyes could also be used to colour food and only tiny amounts were needed. When they were eventually tested on animals many proved to be biologically active and clearly unsuitable for use in foods, but some were safe. Today only those which pass all tests for safety can be used and now we have just a few approved ones such as allura red (E129), carmoisine (E122), ponceau 4R (E124), quinoline yellow (E104), sunset yellow (E110), and tartrazine (E102).

Food colorants have to take into account a variety of conditions under which they will be used and so have to display a range of properties. Are they water soluble or oil soluble? Will they withstand heat when processed? Will they fade when exposed to light? How will they perform at different acidities? (The chemical cyanidin is the natural colorant found in dark fruits like red grapes, and vegetables like beetroot and red cabbage, and it changes colour from bright red to purple depending on the pH.) Cyanidin is an anthrocyanin and there are many versions of this molecule, giving a range of colours. Natural chemical dyes like these are not blamed for causing hyperactivity of course.

Various studies had been made purporting to prove that chemical colorants cause hyperactivity but all were scientifically suspect. Better research was needed, and agitated for by NGOs, so the UK's Food Standards Agency agreed to fund more research at the University of Southampton in 2007. The researchers there decided to test six permitted colours on 3-year-old children and 9-year-old children. They were given one of two drinks: one contained the colours sunset yellow, carmoisine, tartrazine and ponceau 4R, while the other contained carmoisine, quinoline yellow and allura red. To both mixtures they added the preservative sodium benzoate which was also being campaigned against. The colours were added to naturally dark coloured liquids, such as beetroot juice, blackcurrant juice or prune juice, which apparently the children happily drank. Their behaviour was then monitored and indeed these drinks were shown to be affecting the children's behaviour, the 9-year-olds more than the 3-year-olds. The findings were published in a peer-reviewed journal. The NGOs who had agitated for the research claimed its findings supported their campaign for a total ban.

The public relations teams of the NGOs had already primed the media with press releases which duly appeared. "Danger to children from food and drink additives is exposed" was a typical headline. The campaigners were now armed with apparently sound scientific data and off they went to the European Parliament in Brussels convinced that they could get the additives banned.[5] There the European Food Safety Authority passed the report to a committee of 21 scientists who studied the data and concluded differently. The Southampton study had failed to demonstrate a link between the additives and hyperactivity. Its findings had been compromised by confounding variables, not least of which was the

[5] The EU alone has the power to pass laws concerning food additives.

self-selection of parents taking part in the survey, some of whom were even allowed to collect the evidence about the effects these strange drinks had on their children's behaviour. In fact these colorants have now been largely phased out anyway.

You may be surprised to learn that all natural colorants in foods also have to be approved and tested in order to acquire an E number. Carotene, the red colour of tomatoes and carrots is E610; chlorophyll, which colours green vegetables, is E140 (and this is not permitted in the USA); red cochineal (from crushed scale insects) is E120; yellow carotene is E160a; red annatto (from the *Bixa orellana* tree) is E160b; yellow xanthophyll (from various plants) is E161b, beetroot red is E162; and various anthrocyanins are coded E163a, E163b, *etc.* The food industry's response to all the adverse publicity surrounding food colorants has been to try and replace them with these. What has prevented the widespread use of natural colorants in the past is that they tend not to be very stable, more colorant has to be used to achieve the desired intensity of colour, and their colour may change according to the food or drink to which they are added. These nature-identical colorants are not necessarily extracted from the plants which produce them, of course, but are manufactured by chemists.

1.4 NEUTRACEUTICALS AND FUNCTIONAL FOODS[6]

Can natural chemicals in foods protect us from disease?

In addition to the nutrients we require from the food we eat, there are also many other chemicals in food which our body generally treats as unwanted and disposes of *via* our urine and faeces. Some chemicals are highly poisonous and these natural toxins we will come to in a minute. Others appear to be harmless and might even be beneficial. It is this category that we will now consider. A lot has been written about the supposed benefits of certain foodstuffs and indeed there is a health food industry which is geared to providing them – and which generates a lot of favourable publicity in support of them. How much credence can we give to what is being claimed? In this section we will discover that much of what we are told is more fiction than fact.

Neutraceuticals, also referred to as 'functional foods', have become popular in developed countries and are consumed on a large scale.

[6] A food can be said to be 'functional' if it has beneficial effects on the body in addition to its nutritional benefits. The term does not apply to dietary supplements.

However, the functional food bandwagon is becoming stuck in the mud of doubt, despite the scientific evidence it claims in support.

In 1994 it was said that eating unripe bananas could help prevent colon cancer because these contained neutraceutical-resistant starch; in ripe bananas this starch has been converted to sugars. Resistant starch is only digestible by the bacteria living in the colon and as they feed on it they produce butanoic acid, which was known to suppress the growth of cancer cells – at least in the laboratory. Epidemiology supported this theory about resistant starch, witness the low incidence of colon cancer in China and India where starch is a major part of the diet, and the high incidence of colon cancer in North America where starch intake is comparatively low. Point proven? Not really. There are too many confounding variables and no one nowadays supposes that eating unripe bananas will prevent cancer.

Interest in functional foods started in Japan decades ago and they are still very popular in that country. There they distinguish various types of neutraceuticals such as fibre, **polyunsaturated fatty acids**, oligosaccharides and phenols. The first of these, fibre, is undoubtedly important. Fibre is indigestible carbohydrate, mainly cellulose, of which foods like oats, bran, apricots and beans contain a lot. Fibre retains water by virtue of its chemical nature and this aids passage of food through the gut, thereby preventing constipation and some rather embarrassing anal conditions.

There is a lot to be gained by eating at least five portions of fruit and vegetables every day, if only to ensure that you get the complete range of vitamins and minerals your body needs. Some fruits and vegetables also have unique ingredients. Could these be neutraceuticals? Carrots contain falcarinol, a natural pesticide which protects the vegetable against fungal diseases. Falcarinol is said to protect against colon cancer – and that claim is backed up by studies on rats. Cranberries are said to relieve urinary infections. Garlic and onions have high levels of alliin and allicin which protect the heart by lowering high blood pressure and reducing cholesterol levels. Broccoli contains sulforaphane which may boost the type of enzymes that destroy cancer-causing chemicals.[7] Other supposedly functional foods include pomegranates, avocado, beetroot and watercress. Eating more of all of these can only be of benefit even though there is no scientific *proof* that the unusual molecules they contain are truly beneficial.

Two functional food ingredients which have had a good press for decades are **omega-3 fatty acids** and **antioxidants**. However, their days of

[7] Sulforaphane itself cannot be used as an anticancer drug because in large doses it is rather toxic.

glory are coming to an end. For many years I too believed that a daily supplement might help my body ward off cancer and heart disease. Then in 2006 a report in the *Journal of the American Medical Association* concluded that the claims being made for omega-3 were not proven. In the same year a study carried out by a team, led by Lee Hooper of the University of East Anglia in England and published in the *British Medical Journal*, showed that those who regularly took omega-3 food supplements were just as likely to die of heart disease as those who did not take them. And it did not matter whether the omega-3 came from fish oils or from plant oils, both kinds were equally ineffective.

The other neutraceutical components of food that got even more publicity than omega-3 were the antioxidants. Antioxidants were seen as highly beneficial and the existence of such chemicals in certain fruits and vegetables led to these being suggested as ones we should eat more of. Tomatoes contain the antioxidant carotenoid lycopene which was said to prevent cancers of the breast, gut, cervix, bladder, skin and prostate. Carotenoids are the bright yellow, red and orange pigments that give tomatoes, carrots and oranges their colour. They are fat-soluble chemical precursors to vitamin A and are supposed to protect against degenerative diseases by neutralising **free radicals**.

Polyphenol antioxidants were particularly praised and so things rich in these, like green tea and pomegranate juice, became popular. Polyphenols are also found in fruits, vegetables, wines, beers, and teas. Red wine has the polyphenol resveratrol, which is supposed to reduce the risk of heart disease. Tannins are phenolic antioxidants and there are ten times as much of them in red wine than in white wine, although as the following box shows, the evidence in support of red wine is not as convincing as some would have us believe.

Common Sense 3: Red wine protects against heart disease

Maybe. Although several epidemiological studies appear to support this belief they are mainly confounded by the fact that important variables have not been taken into account. For example, red wine is very much a drink of the middle classes and they also have a very different diet and lifestyle to the working class. Middle class people have always tended to live longer in any case. Indeed, the American Heart Association says there has been no direct comparison trial to determine whether drinking red wine has any effect on a person developing heart disease.

In 1981 it was said that 35 per cent of cancers were diet related. It was also postulated that free radicals within the body were the main cause. These highly reactive chemical species could damage the DNA and initiate a cancer, or maybe damage a cell membrane thereby making it possible for toxins and carcinogenic agents to gain access and begin their destructive work. Antioxidants can neutralise free radicals and some antioxidants come in the form of vitamins A, C and E. Dietary studies suggested that eating foods which are rich in these vitamins would protect against cancer and epidemiological studies appeared to bear this out. A study of the diets of 2000 middle-aged men over a 20 year period showed that those who ate fruit and vegetables rich in these vitamins suffered less lung cancer. Other studies appeared to confirm these findings and so the world set sail on a raft of antioxidants.

The antioxidant bubble burst in 2007, or at least that's when medical scientists decided to check whether these substances were as good as was being claimed and if the evidence was trustworthy. The result was not what the supporters of antioxidants were hoping for. *The Journal of the National Cancer Institute* published an editorial by Goran Bjelakovic and Christina Gluud which not only said that taking antioxidant supplements was useless, but that it might even be risky. In 2007 an analysis was carried out of 67 published studies concerning antioxidants. These involved 200 000 individuals, linking the taking of vitamins A and E, and beta-carotene, to general health. The new analysis concluded that these antioxidants had no effect on death rates and that they might even *shorten* the lifespan of those who took them. The 67 studies were chosen as meeting the highest standards, namely involving large numbers of people and with strict guidelines about trying to avoid confounding variables which blighted so many of the other 400 studies in this area.

A report in 2008 from the Centre for Clinical Intervention Research at Copenhagen University Hospital in Denmark also found that taking antioxidant supplements had no effect at all in decreasing death rates, and if anything they tended to have just the opposite effect.

Why should antioxidants be risky? One theory is that the molecules they destroy might in fact be essential for disposing of damaged and unwanted cells from the body, and that it may be unwise to interfere with this defensive mechanism. Such cells might include cancerous and pre-cancerous ones. Support for this idea came from a study of the effects of a daily intake of a multivitamin supplement on 300 000 men who were all free of cancer when the trial started. The finding was that those on the multivitamin regime were more than twice as likely to develop fatal prostate cancer. So much for multivitamins.

Two types of functional food are not designed to provide us with dietary components, but to provide food for the microbes which inhabit our gut. Some of these are good, such as *Bifidobacter* and *Lactobacillus*, and some are bad, such as *Enterobacteriaceae* and *Clostridium*. We can change the balance between the two types by what we eat, namely by consuming so-called *probiotic* and *prebiotic* foods. The former are meant to introduce the good bacteria into our intestines, while the latter feeds those that are already there in the hope of boosting their number and thereby crowding out the bad bacteria, which have been blamed for causing irritable bowel syndrome, heart disease, allergies, autism and even ageing.

In the 1930s Minoru Shirota of the University of Kyoto searched for good bacteria that would survive passage through the acid conditions of the stomach and recolonise the intestines and colon. He cultured samples taken from human faeces, and discovered *Lactobacillus casei Shirota* (named after him), and this he developed as the fermented milk drink Yakult, and which is now sold around the world.[8] Another probiotic drink is Danone's Actimel which has *Lactobacillus casei immunitas* as its active bacteria. With regular consumption of such probiotics the composition of the gut microbes can be changed and if the theory is correct it should benefit one's health. Or maybe not.

A report in *The Lancet* in February 2008 revealed something rather disturbing. In a trial to see if probiotics were a treatment for acute pancreatitis, which is a painful inflammation of the pancreas, doctors found that more than twice as many patients consuming a probiotic drink died than those in a control group. This was a **double-blind test** in which 152 patients were given a probiotic drink twice a day while 144 patients were given only a placebo version. Of the test group 24 died, but only 9 of the placebo group. Nine patients on the probiotic developed bowel ischaemia (restricted blood supply) of which 8 died, but none of the placebo group developed this complication. Equally disturbing has been the discovery that a good bacterium, *Bifidobacterium longum*, can lose some of its genes when cultivated under the conditions used to produce it. The genes it loses are exactly those needed to survive in the hostile environment of the human gut.

If you don't like the idea of consuming live bacteria then the alternative is to feed the good bacteria that you already have with so-called 'prebiotics' and thereby increase their number. Prebiotics are non-digestible carbohydrates known as **oligosaccharides** and they can pass

[8] Each 50 ml bottle of Yakult contains more than 6 million of these bacteria.

through the stomach and the small intestine without being broken down, so that they reach the colon intact there to feed the good bacteria. These carbohydrates may be added to all kinds of foods, such as cereals, cakes, biscuits and health drinks. They can be extracted from things like chicory root or produced from sugar by the action of specific enzymes. They also occur naturally in small amounts in bananas, leeks and wheat. There are reasons to believe that prebiotics might be useful. Indeed breast milk contains several prebiotic carbohydrates which a baby cannot digest and these appear to be there to feed the friendly *bifidobacteria* in the baby's gut. The more of these there are in the gut, the less there will be of those which cause gastrointestinal infections. Mother's milk has between 3 and 15 g per litre of these carbohydrates. Infant formula feeds now contain added prebiotics to make them more like breast milk. It seems more than likely that prebiotics can do no harm.

1.5 NATURAL TOXINS AND NATURAL DETOXING

Because something is natural does not automatically mean it is good for you.

As far as medical science is aware, there is no evidence that minute traces of pesticides in food present a danger to health. The same, however, cannot be said of natural toxins which plants make in order to protect themselves. In this section, I want to look at five natural molecules which are present in sufficient quantities to affect those who eat the plants which produce them. These are: linamarin, oxalic acid, salicylate, solanine and the pyrrolizidine alkaloids.

Linamarin. Cassava root is a carbohydrate-rich staple food eaten in many countries, especially in Africa. Around 500 million people rely on it, some people eating half a kilogram a day. It is consumed as a kind of porridge made by peeling the root; grating it; soaking the pieces in water for three to five days; straining off the water; drying the cassava in the sun; then grinding it into flour with which to make the porridge. Cassava contains linamarin and this has a cyanide group as part of its molecular makeup which this lengthy method of preparation removes. Eating cassava that has been prepared without due care will lead to symptoms, such as paralysis of the legs, blurred speech, dizziness, headaches and vomiting. (In theory there is a fatal dose of cyanide in one cassava root.) Grating the root releases an enzyme which breaks down the linamarin and releases the cyanide as hydrogen cyanide (HCN), which dissolves in the water and so is discarded, or which evaporates when the cassava is dried.

Oxalic acid. In times of food shortages people have been known to cook rhubarb leaves as a vegetable and become very ill as a result, suffering from oxalic acid poisoning. The acid or its compounds, called oxalates, are present in chocolate, peanuts, spinach, parsley, beetroot and tea, and in large amounts in sorrel. Soup made from this has caused at least one fatal case of poisoning. Too much oxalate causes sore throat, abdominal pain, diarrhoea and occasionally vomiting of blood. Both the liver and kidneys are damaged and may take weeks to recover. The foods with the highest oxalate content should be avoided if you are at risk from kidney stones which are mainly calcium oxalate. Other foods containing oxalate include: runner beans, broad beans, baked beans, celery and strawberries.

Salicylate. Lots of foods contain salicylate and those who are affected by this simple molecule need to avoid things like currants and raisins, oranges and raspberries, almonds and tea. Nor must they take aspirin, which is a derivative of salicylate, and while this has beneficial effects for most people, a few are highly sensitive to it and may suffer stomach bleeding. Salicylate-rich foods can provoke an intolerance response in the form of a headache, vomiting and diarrhoea. It has this effect because it interferes with certain key enzymes. Those who are ultra-sensitive to salicylate will experience symptoms such as urticaria (aka hives), swollen lips, swollen tongue, difficulty in breathing, upset stomach, irregular bowel motions and a general run-down feeling. Only a trained dietitian can advise on a salicylate-free diet because this toxin is found in so many foods, including apples and honey – herbs and spices contain a lot – but there are foods which contain none, such as bananas, rice, bread, cheese, yogurt, eggs, beef, chicken and tuna.

Solanine. This is present in small amounts in all potatoes but the level increases to dangerous levels when they are exposed to sunlight, or when they have started to sprout. Solanine is thought to protect the tubers, and it can have a toxic effect when such potatoes are eaten. Accidental poisoning by potatoes is not uncommon, and in 1984 a report was published of 12 well-documented incidents involving 2000 people, in which there were many hospitalisations and 30 deaths. These generally involved institutional catering because it is only when there are several simultaneous cases that the incident is officially noted. Domestic cases tend to go unreported or are blamed on other illnesses, because the symptoms of nausea, vomiting, diarrhoea, stomach pains, headaches and dizziness are generally associated with food-poisoning caused by microbes.

Pyrrolizidine alkaloids. Herbs contain natural toxins but some are so dangerous that they have been banned, such as the herb comfrey. As part of alternative medicine therapies, comfrey was advocated as a treatment for arthritis, headaches, common colds and even as a way of preventing cancer. Its leaves could be eaten as part of a side salad, or infused to produce comfrey tea, or concentrated into a capsule of comfrey extract. Yet eating just a few comfrey leaves is enough to cause sickness and vomiting in some people because they contain pyrrolizidine alkaloids (PAs). These substances cause cirrhosis of the liver by blocking veins in the central lobe. According to the German Health Authority, even small doses of PAs can affect genetic material, harm embryos and even cause cancer. Grazing animals may sometimes eat enough PAs from the weed ragwort to kill them.

If you are one of those people who imagine that their body is accumulating too many 'chemicals', then you may be persuaded to buy a detox treatment. There is no scientific evidence to substantiate these and they are of little benefit.

Common Sense 4: I need to detox my system to remove dangerous chemicals

Wrong. Patent detox treatments are big business – and a waste of money. The London-based group Sense About Science has published a booklet, which can be downloaded from their website at http://www.senseaboutscience.org.uk/pdf/MakingSenseofChemicalStories.pdf and is specifically designed to counter the nonsense talked about detox methods and products. In fact we come into the world with a wonderful natural detox system called the liver, whose job it is to filter out of the blood those molecules for which the body has no use, either turning them into something which can be used, for example alcohol can be converted to acetic acid and thereby supply energy, or simply passing them on to the urine for disposal. If you feel the need to detox yourself, say after the excesses of Christmas and the New Year celebrations, then all you need is to eat less, drink more water, and go to bed early. After three days of such a regime your body will be detoxed – and it costs nothing.

1.6 FOOD FRAUD AND FOOD FORENSICS

Food adulteration is an age-old problem but chemical analysis can defeat it.

Food adulteration costs the consumer around $50 billion a year globally; the UK Food Standards Agency estimates it to be around $5 billion in Britain.

Caveat emptor! Buyer beware! When food became something you bought rather than produced yourself then you were always at risk of fraud. That was true in Babylon in 1750 BC where laws written on clay tablets reveal that watered down beer was a problem. Science was first used in the 1850s to prove that ground coffee was being adulterated. When examined by microscope, it could be seen that samples had been adulterated with chicory, roasted wheat or burnt sugar. Over the years coffee has been adulterated by other roasted plant materials but these give themselves away by the presence of things like dextrins and fructose which should not be there. When the UK Ministry of Agriculture, Fisheries and Food analysed instant coffees in 1993 they found that 15% of them were adulterated. Even now coffee adulteration has not stopped. Arabica coffee is considered to produce the best drink, but more profit can be made by adding the much cheaper Robusta coffee. That fraud can be exposed by analysing for the natural chemicals: chlorogenic acid and caffeine, the ratios of which are different in the two coffees.

In the 1970s it was common for honey to be adulterated with high-fructose corn syrup until food chemists were able to show that this form of cheating was widespread. In the 1980s the great scandal was Austrian wine to which unscrupulous vintners were adding ethylene glycol to give it more body and thereby charge a higher price. In the 1990s orange juice was adulterated with glucose, sucrose and fructose which came from **invert sugar**. This adulteration went almost undetected because orange juice contains these same carbohydrates in the same proportions. However, the addition of these also required the addition of more fruit acid, and the one chosen was **malic acid** which is also naturally present in fruit juice. This exists in two forms, a left-handed (L) and a right-handed (D) form, of which natural orange juice contains only the left-handed form. However, the synthetic form of malic acid is a mixture of the two and the presence of D-malic acid in the orange juice exposed the fraudsters.

Those who adulterate food are now almost certain to be caught even when the degree of adulteration is at incredibly low levels, as it was in the Sudan-1 scandal of 2005. This is a deep red dye and it is not permitted as a food colorant because it is carcinogenic. However, it had found its way into ground paprika produced in India and Uzbekistan and then into things like Worcestershire sauce and cayenne pepper. The amounts of Sudan-1 reaching the consumer in the form of curry meals and Mexican dishes was at levels which presented no threat, but it was enough to cause supermarket shelves to be emptied of such products and tonnes of food to be destroyed.

Figure 1.3 Chemists were able to prove that a lot of orange juice was not the real thing.

In China, in the summer of 2008, more than 300 000 babies were affected by contaminated formula milk powder: of these 6000 became seriously ill, 150 suffered kidney failure and 6 died. In fact they were being poisoned by the chemical **melamine** which was being added to milk powder to make it appear to be more nutritious that it really was. The protein content of milk is simply assessed by analysing for nitrogen. Protein consists of amino acids and amino acids have a nitrogen atom as an essential part of their molecular structure. Analyse milk for its nitrogen contents and this reveals its protein content, which is an all-important ingredient for babies.

Adding a nitrogen-rich molecule to milk can make it appear better than it really is and melamine was used for this purpose. Melamine has the chemical formula $C_3H_6N_6$ and it is 67% nitrogen, so that adding a little of this to milk can boost its nitrogen analysis. Such milk, produced by four dealers, was being processed at more than 20 companies and the scam had been going on for quite some time. Zhang Yujun, 40, was brought to trial in December 2008 charged with making around 600 tonnes of melamine specifically to add to milk. He was sentenced to death, as were Geng Jinping and Gao Junjie who added the melamine to the formula milk. Tian Wenhua, 66, the chairwoman of the Sanlu Group which sold the product, was sentenced to life imprisonment. The scam came to light when a New Zealand company which imported the milk powder revealed what was going on.

Milk powder was also exported from China and used to make products like chocolate, biscuits and toffee, and alarms were ringing all round the globe. The European Food Safety Authority issued a reassurance that children were not at risk provided they did not eat excessive amounts of such foods. In fact the human body can tolerate low levels of melamine and the World Health Organisation (WHO) have said that 0.2 mg per kg of body mass per day is tolerable. In future it will be possible to detect melamine very quickly using a technique known as **electrospray ionisation mass spectrometry** which will identify the adulterant immediately. Other quick methods of measuring its presence have been developed and Chinese chemists now have a simple test based on the chemical luminal which emits a blue glow if the adulterant is detected.

The leading food analysts in the UK are RSSL (Reading Scientific Services Limited) and they use a method for determining melamine called ELISA, which stands for **e**nzyme-**l**inked **i**mmuno**s**orbent **a**ssay and which relies on antibodies to detect the adulterant and signal this with a fluorescent marker.

A similar melamine scandal occurred in the US in 2007 when melamine-adulterated pet food killed as many as 1000 cats and dogs. Again the source was ingredients imported from China which had been adulterated so as to make them appear to have more protein.

The WHO analysed the melamine produced in China and found it to be fairly pure, and as this chemical is tolerated fairly well by the human body they sought another reason why it proved so deadly to some babies. The answer appears to be its interaction with uric acid, which the body produces as it seeks to rid itself of excess nitrogen and which is particularly high in the urine of babies. Melamine and uric acid form crystals in the liver and this caused liver failure.

While melamine in food might be easily measured because it is an unnatural ingredient, it is less easy to prove that virgin olive oil has been diluted with a cheaper olive oil, or that cod liver oil has had rapeseed oil added to it. The latter fraud was being perpetrated in the UK, and its came to the attention of the authorities when it was noticed that the amount of cod liver oil leaving a processing plant was greater than that entering it. However, forensic analysis of its fatty acid content conformed to those of genuine cod liver oil. The answer lay in the traces of sterols in the oil and these come from the plankton which the fish eat. These live in the Arctic and their sterols can be shown to be present in the oil of its liver. They are not the same sterols as those in rapeseed oil. The analysts calculated that the adulterated cod liver oil contained 12% rapeseed oil.

And what about olive oil? How do you know that really expensive extra virgin olive oil, which supposedly comes from one particular olive grove, is pure? The answer lies in the DNA of the cultivars of trees in that orchard. Only traces of that should be in the oil. Any other DNA shows that the product is not as exclusive as is being claimed.

Another product that is easy to adulterate – again in a non-threatening way – is mozzarella cheese. How can you tell whether some cow's milk has been added to the buffalo milk before the cheese is made? When food chemists at the University of Padua analysed 64 samples of buffalo mozzarella they found most of them had been adulterated and they proved this by using DNA analysis which detected genes specific to cow's milk. (And when mozzarella is used in pizzas then the tendency is to replace some of it with a cheese substitute made from vegetable fat.)

Methods for forensic food analysis can be even more sophisticated. Corn-fed chicken, *i.e.* chicken fed exclusively on maize, commands a higher price because this is a more expensive feed than wheat and barley. Proving that these cheaper feeds have been used is possible by measuring the ratio of the isomers ^{13}C and ^{12}C in the chicken flesh. This ratio differs in maize because this plant has a different metabolism from wheat and barley.

There is no way today of hiding food adulteration, thanks to the advances in chemical analysis, but there are no doubt other frauds still being perpetrated which would be difficult for the consumer to detect. Could normal food even be passed off as 'organic' food and priced accordingly? Very easily, as was proved in 2009. Neil Stansfield had an organic food company called ONE food (**O**rganic, **N**atural, **E**thical) whose Saddles brand was even on sale in London's up-market food store Fortnum & Masons. It was a multi-million pound scam, because all Stansfield did was buy cheap supermarket food and repackage it in expensive Swaddles wrapping. His scam went unnoticed for almost five years; his prison sentence was half as long.

In this chapter, I have maintained that organic food has few advantages to offer in terms of nutrition or health over conventionally grown food, and there is no way that a world population of billions could be sustainable if it were to rely on organic food. I hope I have convinced you.

Common Sense 5: 'Organic' food is healthier

Wrong. This is the reason many people are lured into paying premium prices for 'organic' produce but they are wrong in their belief says a review funded by the Food Standards Agency (FSA), and headed by Dr Alan Dangour of the London School of Hygiene and Tropical Medicine. The paper, published in *The American Journal of Clinical Nutrition* in July 2009, reported that the studies carried out over the past 50 years which claimed 'organic' food to be superior were unconvincing.

Sometimes 'organic' food can be very unhealthy, even deadly. In the summer of 2008 in California, 204 people became ill through eating 'organic' spinach. Half of them were hospitalised, 30 suffered kidney failure and three died, including a two-year-old child. The cause was the deadly bacterium *E. Coli* 0157:H7. By analysing the DNA of this microbe it was possible to trace its origin back to the source, a company called Natural Selection Foods, which purchased its spinach from an 'organic' farm where free range pigs manured the field where the spinach had been grown.

CHAPTER 2

Water and Chemistry

[A word in **bold** means there is more information in the Glossary.]

Water is the limitless resource that's in short supply. In this chapter, we will look at seven aspects of water in which chemistry has a role to play. They are: drinking water; water for washing-up; water analysis; wastewater treatment; water for irrigation; water from seawater; and extremely pure water.

Ever year more than 5 million people die, most of them children, from water-borne diseases. To highlight the problem, the UN has designated 22 March of every year to be World Water Day, and in 2005 that day was also chosen as the start of the Water for Life Decade. A similar decade in the 1980s brought clean water to 1 billion people and sanitation to around 100 million. Yet there are still around 1 billion people for whom clean water is not available at the turn of a tap, and 2 billion who lack basic sanitation. The target now is to halve these figures by 2015. Chemistry's role will be vital.

The minimum amount of water that a person needs every day for drinking, cooking and washing is 20 litres, although life is clearly much better if double or triple that amount is available. Our daily intake of water is around 2 litres, of which about a quarter comes from the food we eat. The water content of food varies widely, for example, an egg is 75% water, fish is 80%, an orange 85%, carrot 90%, and lettuce 95%. Even cheese is 35% water and butter 15%, while peanuts have only 5% and a chocolate biscuit has almost none at all. During the course of a day the average human sweats 500 ml of water, breathes out 500 ml of

A Healthy, Wealthy, Sustainable World
By John Emsley

Published by the Royal Society of Chemistry, www.rsc.org

water vapour, and urinates 1000 ml of water, a total of 2 litres. These losses of water can double if a person engages in strenuous activity.

Drinking water is only a small fraction of the water we require. The average home takes in around 400 litres of water a day for toilet, washing and cleaning but this is not the only water we rely on. Eat a hamburger and drink a milkshake and demand now soars, because it takes more than 3000 litres of water to produce a litre of milk and around 10 000 litres to produce a hamburger – and even that water may not represent the real total.

In 2008 the Stockholm Water Prize was awarded to Professor John Anthony Allan of King's College London[1] in recognition of his theory of 'virtual' water. This is based on goods imported into one country which require a large volume of water to be used in their country of origin. Thus the UK, which has a plentiful supply of water, actually has a large virtual water footprint because of the food and cotton it imports. Allan's theory suggests that 60% of the UK's water consumption is virtual water. Buy 1 kg of imported beef and your water footprint is 15 000 litres; buy a pair of foreign leather shoes and it is 8 000; a cotton shirt is 2 700; a cup of coffee is 140; and an apple is 40. Even a single sheet of A4 paper consumes 10 litres.

Water is needed for agriculture, industry and homes. A lot is wasted, a lot becomes polluted and only a little is recycled. The Earth is literally awash with water and the numbers are staggering. There are around 14 000 000 *trillion* cubic metres (m^3) of it and we know that it covers 70% of the globe, and it equates to 9 billion m^3 per human being. However, of this total less than 1% is accessible as freshwater in lakes and inland seas. There is also non-salt water in the ground, in ice caps, in glaciers and in the atmosphere.

Because water is volatile, it has been endlessly evaporating, condensing and dissolving rocks for billions of years so that most of it now exists as a saline solution. While some life forms have adapted to living in salt water, those which inhabit the land have not. They need fresh water and rely on the 100 trillion tonnes of rain which fall every year, and while this seems a lot, rain tends to fall heavily in some areas and hardly at all in others. Fresh water has to meet several needs and an important one is agriculture, which consumes around 70%. If savings could be made in this area, it would release fresh water for other uses where it is badly needed, and it could relieve some of the

[1] This prize of $150 000 was instituted in 1991.

water shortages which afflict certain areas – and that includes even the USA.[2]

The UK Government's Department for Communities and Local Government has issued a Code for Sustainable Homes which explains how new homes are to be graded with respect to various factors including water use. The less water consumed per person in a home, the higher the rating, so that if the average consumption is 120 litres per person per day then the house is rated 1 or 2, if it is 105 litres then it is rated 3 or 4, and if it is 80 litres then the rating is 5 or 6. There is more on sustainable homes in Chapter 6.

2.1 DRINKING WATER

Making water fit to drink requires just simple chemistry, but this has led to controversy.

Common Sense 6: Bottled water is healthier

Wrong. Bottled water offers no health benefits and indeed is bad for the planet because it wastes energy spent collecting, bottling and transporting it from sources such as springs or glaciers. In some cases, it is being transported more than ten thousand miles. Take Fiji water for example. It makes the following claims:

> "*Far from pollution. Far from acid rain. Far from industrial waste. There's no question about it: Fiji is far away. But when it comes to drinking water, 'remote' happens to be very, very good. Look at it this way. Fiji Water is drawn from an artesian aquifer, located at the very edge of a primitive rainforest, hundreds of miles away from the nearest continent. That very distance is part of what makes it so much more pure and so much healthier than other bottled waters*".

While some of this public relations blurb may be true, its claim to be more pure and healthier than other bottled waters is nonsense. In fact Fiji water, in common with most bottled water, will damage your health by wasting money you could be spending on things which might really be helpful, like fresh fruit and vegetables.

[2] In the US there is a demand for 1.3 trillion litres of water a day, of which 40% is used for irrigation; 39% for power generation in the form of water used in cooling systems; 13% for domestic use; 5% for industrial use mainly in refineries, steel mills, pulp mills and the chemical industry; and 3% for other uses such as mining.

Figure 2.1 Everybody benefits if it's water from the kitchen tap.

Almost certainly the readers of this book will enjoy the benefits of a modern water industry which is based on 150 years of improvements. These began in 1850s London, then the world's largest city, and they were triggered by the way that drinking water was sourced and foul water was disposed of. In 1854 John Snow proved that an outbreak of cholera was linked to a well in the Soho district which was being contaminated with human excrement. Four years later, in a heat wave, the river Thames caused 'The Great Stink' because that was where London's sewers discharged their untreated contents. The smell was so bad that the Houses of Parliament suspended business. What London needed was a supply of safe water and a sewage treatment system. By the end of that century it had both, thanks in part to the developing science of chemistry.

There are around 150 000 water treatment plants across the world and these rely on chemicals to make the water free of dirt, impurities and water-borne diseases, such as cholera, typhoid, dysentery and meningitis. Impurities in water are of four kinds: microbes, particulates, inorganics and organics. Microbes can be bacteria, viruses and protozoa; particulates can be silt and clay; inorganics can be toxic metals; and organics can be plant debris.

Chemistry is the key to producing clean water. The steps involved are flocculation, sedimentation (or flotation), filtration and disinfection. Some of these have been improved in recent years, driven by the awareness that what was thought to be clean water 30 years ago actually contained things which we would rather not have present. These micro-pollutants are pesticides, organochlorine molecules and trace metals. The first of these were mainly the herbicides used by local authorities for weed control or the pesticides used by farmers to protect crops. The second were formed from dissolved organic matter in the water, most of which is the humic acid of decaying leaves. The third group are metals, some of which come from industry, such as nickel, some from water pipes, such as lead, and some from natural sources, such as arsenic and uranium. By the end of the last century water regulations were stipulating the maximum permissible levels of all these unwanted substances, so that less than a **microgram** a day would be consumed by those drinking it.[3]

Before we need worry about these micro-pollutants, we have the macro-pollutants to remove. When water is drawn from a lake, river or well, it may be cloudy and so the first process is to clarify it, although if it is suspected of being highly contaminated with microbes then the first step may be to bubble **ozone** through it to kill them. Following this process, comes the flocculation and settling stage. No one wants to drink cloudy water and cleaning it up is usually done with **coagulants** of which aluminium sulfate and iron(III) chloride are typical. When solutions containing these are dripped into cloudy water, made slightly alkaline with calcium hydroxide (aka hydrated lime which has the chemical formula $Ca(OH)_2$), they form a flocculent precipitate (the 'floc') which settles out. The floc is a coagulation of tiny particles into millimetre-sized ones, and these will absorb most of the algae, bacteria and viruses, as well as the dissolved organic matter. The floc can fall to the bottom of the tank under the action of gravity (settlement) or be raised to the top of the water by the action of a stream of fine bubbles of air (flotation). Either way it can be removed. The water can then be filtered through sand and activated carbon to remove residual traces and is now crystal clear.

Water needs to be disinfected before it is allowed to enter the mains and this is generally done with either **ozone** or **sodium hypochlorite**

[3] This would amount to a mere 27 mg in a lifetime of 75 years.

(aka bleach) to destroy any disease-causing viruses and bacteria which might remain. Destroying these is part of the battle we wage to remain healthy, and the best ammunition is sodium hypochlorite which has probably saved more lives than any other chemical because it is deadly to all micro-organisms and it has been part of water treatment for more than a century. Hypochlorite keeps drinking water free of disease pathogens, and not only of the deadly ones, but the more common ones like *E. coli* that cause vomiting and diarrhoea. It also ensures the water remains germ-free until it reaches the end users.

Bleach was first used to disinfect tap water at Maidstone, England, in 1897 to bring an outbreak of typhoid under control. Its efficacy was confirmed when it stopped another epidemic at Lincoln a few years later. Eventually it became the method of purifying drinking water throughout the British Isles and eventually throughout most of the developed world. Sodium hypochlorite is made from chlorine gas, which is manufactured at chemical plants and then either shipped as the liquefied gas in tankers, or converted to the final product sodium hypochlorite before being shipped. Chlorine will always be a sustainable resource for the chemical industry because it is made from salt which can be extracted from sea water.

Shipping liquid chlorine always has the potential to cause a terrible accident, as happened in the US on 6 January 2005 when a rail tanker carrying 60 tonnes broke open after it was derailed in a crash. The resulting cloud of chlorine vapour killed 9 people and injured hundreds of others in the town of Graniteville in South Carolina. Shipping sodium hypochlorite solution, which is much less risky, is nevertheless a waste of energy because more liquid has to be transported. Consequently, water companies are now turning to technology which generates hypochlorite on site from salt solution by means of electricity. Severn Trent Services have been selling such generators since 1989 and there are now more than 3500 of their installations in use and these produce a 1% solution of hypochlorite, ideal for water disinfection.

Bleach kills germs by means of the hypochlorous acid (HOCl) it forms in solution, and this is safe at the levels used because it is the chemical which our own immune system produces to destroy the germs which invade our body. In 1996, it was reported in *The Journal of Clinical Investigation* that white blood cells produce hypochlorite to fight off invading microbes and they rely on the enzyme *myeloperoxidase* to make this chemical for them. Like so many chemical discoveries that benefit us, Nature did it first.[4]

[4] If you find traces of hypochlorite not to your taste then you can buy filter jugs which will remove it.

The disinfection of water using UV is also a possibility and this may one day be the method of choice. It has some obvious advantages: it inactivates bacteria within seconds; destroys viruses; adds nothing to the water; requires no large retention tanks; is safe to operate; and it leaves no undesirable by-products. However, it is complicated and expensive and does not provide residual disinfection.

Do-it-yourself water purification

The chemists at Procter & Gamble, which manufacture household chemicals, have devised a way of making dirty water fit to drink. The product, called PüR, comes in a sachet, costs only a few pence and more than 50 million have been sold in countries like Guatemala, Pakistan, Bangladesh and Uganda. It consists of 4 g of two chemicals which, when stirred in 10 litres of water, will render it fit to drink. These chemicals are iron sulfate and calcium hypochlorite. The iron sulfate produces a floc of iron hydroxide which absorbs things like trace metal contaminants and organic matter which can then be filtered off, while the calcium hypochlorite kills all the germs. Where PüR is available it has reduced the incidence of diseases like diarrhoea quite dramatically.

Another method of water purification is called SODIS (solar water disinfection) and it has the approval of the World Health Organisation. It involves nothing more than a clear plastic bottle into which the water, which must not be cloudy, is left in the sun for a few hours. This kills around 99.99% of pathogens making it fit to drink, although of course it does not remove things like toxic metals.

A quicker method of disinfecting drinking water is to use sodium **dichloroisocyanurate** (NaDCC) which is sold in packs for travellers to regions where the quality of drinking water cannot be guaranteed. Its taste may be somewhat unpleasant but at least it makes the water safe to drink. A typical tablet contains 8.5 mg of NaDCC and this will sterilise a litre of water. It forms a hypochlorite solution strong enough to kill almost all pathogens but is not too strong to make the water undrinkable.

'Chlorine is the Devil's element' was the slogan Greenpeace used 20 years ago in its campaign against bleach, claiming that it formed organochlorine molecules, such as chloroform, and that these might cause cancer. They helped convince the Peruvian authorities to stop treating their water this way in 1990 and the result was an outbreak of cholera with a million cases of the disease, of whom 12 000 died. Needless to say, few other countries followed their example, although some turned to ozone as an alternative water disinfectant. The epidemiological data which underpinned the campaign against bleach was far from convincing. The most common organochlorine residue in water is chloroform, but even at 100 parts per billion (which is equivalent to a tenth of a milligram per litre) you would consume only 3 grams of chloroform in a *lifetime* from this source.[5]

In 1991 the International Agency for Research on Cancer (IARC), which is part of the World Health Organisation, evaluated the risks posed by organochlorines in drinking water and its report said that if there were any health risks they were very low, and they had to be offset against the much greater health risks associated with drinking unchlorinated water. Unfortunately, this cautionary note came too late to save lives in Peru.

Disinfecting water can be done with sodium hypochlorite (NaOCl), monochloramine (NH_2Cl), chlorine dioxide (ClO_2) or ozone (O_3). Which chemical is used depends on the country concerned, for example, the USA prefers monochloramine, whereas this is forbidden in some countries of the EU. The trouble with ozone is that while it does kill pathogens as effectively as hypochlorite, it does not linger long enough in the water to continue this protection it in its journey to the consumer.

Chlorinated drinking water may still contain one particularly nasty individual, which protects itself with a coating that hypochlorite will only destroy at levels which some people find unpalatable. These microbes are eggs of the parasite *Cryptosporidium,* which are shed in the faeces of animals and can find their way into surface waters and drinking supplies. People infected with this parasite have severe stomach cramps and watery diarrhoea lasting for two weeks and in some cases the condition can be severe enough to threaten life. In 1993 in Milwaukee, Wisconsin, there was an outbreak which affected more than 400 000 people, and 100 died.

In addition to hypochlorite, other chemicals might be added to the water, such as hydrofluorosilicic acid which helps prevent tooth decay; phosphoric acid to prevent lead and copper pipes from leaching these toxic metals into the supply; and maybe even ammonia which converts hypochlorite to chloramine (NH_2Cl), which is longer lasting than

[5]This amount should be compared to the chloroform that was once available as the over-the-counter medicine Chlorodyne, a patent cure-all which contained 14% chloroform. A single dose of this provided 3 grams of chloroform.

hypochlorite and is less likely to react with organic compounds to form organochlorines.

It is generally acknowledged that a little fluoride in drinking water strengthens tooth enamel and prevents decay, and to this end many public water supplies are fluoridated to a level of 1 **ppm** (1 mg per litre). There are dangers in drinking water that has much more than this, and in some localities the fluoride levels are naturally high. As a result, humans can suffer from fluorosis, which is a hardening of the bones leading to a deformed skeleton. In certain parts of India, such as the Punjab, the condition is endemic, especially where some well water has 15 ppm of fluoride. About 25 million Indians suffer a mild form of fluorosis, with thousands showing skeletal deformities.

2.2 LET CHEMISTS DO THE WASHING-UP

A lot of water can be saved when it comes to doing the washing-up.

The end of a family meal is followed by one of the least loved chores of everyday life: the washing-up. This can be done by hand or by machine. The first method uses lots of hot water, a liquid detergent and human energy; the second uses less hot water, a solid detergent and electrical energy. Not surprisingly the latter method is becoming increasingly popular and we might think that this must be less environmentally friendly than washing-up by hand. But is it? Box 7 shows how the traditional method uses almost four times as much water.

Dishwasher detergents need to remove all kinds of food residues, such as fats, starches, proteins and stains like those of tea, wine and fruit juices, from all kinds of surfaces, such as crockery, plastic, glass and metal. Over the years, dishwasher chemicals have improved, going from a simple powder to two-in-one tablets, three-in-one tablets, four-in-one tablets, and even five-in-one tablets.[6] They are tested on residues which are known to be the most difficult to remove, namely tea stains, baked lasagne, boiled milk, dried-on porridge and scrambled egg. Test samples are cooked in glass dishes, apart from tea stains for which white cups are used. To make the test more realistic, the stains and residues are left for 24 hours before being washed.

Most people in the UK live in hard water areas and this is why a water softener has to be included. The ingredient which does this is penta-

[6] In fact there are not 5, but 25 ingredients in some tablets.

Common Sense 7: It's greener to do the washing-up by hand

Wrong. In fact, dishwashers save water, energy and time. In 2008 I was an independent observer of some research to test this, involving 150 people selected on the basis of their diverse origins representing all parts of the UK. They were asked to wash-up by hand a load that a normal dishwasher can cope with (equivalent to 12 place settings). They were provided with everything they needed. The amount of energy used to heat the washing-up water and its volume were exactly measured, as was the time they spent. The results are given in the table below and you can see that the dishwasher won hands down.[7] In general, women used less water than men and got things cleaner although still not as clean as a dishwasher. In March 2009 the UK's Consumer Association endorsed dishwashers as more energy and water efficient than washing-up by hand.

	Water used (litres)	*Electricity used (kWh)*	*Time taken (minutes)*	*Cleaning score**
Machine (Bosch)	13	1.3	10**	4.2
People (average of 150)	50	2.0	30***	3.7

*Score 5 = perfectly clean; 4 = a few water marks; 3 = several water marks; 2 = lots of water marks; 1 = some soiled areas remain; 0 = lots of soiled areas remain.
**Time taken to load and unload a dishwasher.
***This is the minimum time needed to wash a 12 place setting. Some people took more than an hour.

sodium triphosphate and it is there to combine with any calcium in the wash water and keep it from forming limescale ($CaCO_3$).

The next most important component is sodium percarbonate, a stable solid made from sodium carbonate (Na_2CO_3), and **hydrogen peroxide** (H_2O_2). This removes tea, coffee, red wine and fruit juice stains. To work effectively it needs the activator **TAED**. The hydrogen peroxide reacts with TAED to form perethanoic acid, and it is this which oxidises and decolourises the stains.

Metal items such as silverware can be badly affected by the bleaching agent and they too need protection, which is provided by 1,2,3-benzotriazole. This is attracted to silver and coats the metal with a layer one molecule thick and so safeguards it from corrosion.

To remove fats and oils we need **surfactants**. These molecules consist of a water-seeking head attached to an oil-seeking tail and they encircle the

[7] Reckitt Benckiser, who manufactures dishwasher tablets, paid for the research which was carried out by an independent testing organisation.

grease and carry it away. Washing-up liquids contain so-called amphoteric surfactants which work well for all kinds of surfaces, are gentle to the skin and they foam easily. However, foam is not something we want in a dishwasher because it blocks pipes. For dishwashers, a non-ionic surfactant is used, such as a **fatty alcohol ethoxylate**, and especially one in which the hydrocarbon chain of the fatty alcohol has 12–14 carbon atoms. These surfactants clean well and produce almost no foam.

Other components of dishwasher formulations are protease and amylase enzymes which are there to break down protein and starch food residues into their water-soluble components. There is also sodium carbonate to make the wash water slightly alkaline which helps in the removal of grease, and zinc acetate (aka zinc ethanolate) to prevent glassware developing an iridescence, although after 50 or more washes this still occurs to some extent.

At the end of the wash process comes the rinse cycle, and here again a little more surfactant is needed. It is not there to do more cleaning but to reduce the surface tension of the water so that it drains off the washed items and does not form droplets, which dry out leaving behind unsightly spots.

The latest versions of dishwash tablets come with various sections sealed inside compartments made of the water-soluble polymer **polyvinyl alcohol**. Some ingredients in a dishwash tablet have to be kept separate to prevent them from reacting chemically with one another. Enzymes should not be in contact with the peroxide bleach for example.

Is it possible that research chemists might one day make dishwash tablets entirely from sustainable resources? The answer is 'yes' and indeed they are already moving in that direction. In many parts of the world phosphate is now extracted from waste waters and recycled as mineral phosphates to provide a sustainable resource. Surfactants can be made from plant materials, and indeed some are already made by linking a carbohydrate, such as glucose, to plant oils, such as coconut oil. The products are known alkyl glycosides and are manufactured at a rate of more than 100 000 tonnes per year. One of these, lauryl diglycoside, is already used in some washing-up liquids.

Within a generation we might well see all the surfactants that currently come from fossil fuels being replaced entirely by chemicals derived from natural materials like plants. And if chemists could make them work just as well in cooler water, say at 40 °C, then they really would be green.[8]

[8] This temperature would not guarantee disinfection, however.

2.3 WATER ANALYSIS

Chemical analysis of tap water is a vital part of quality control and safety.

Chemists have been analysing water for more than two centuries and with increasing skill, so that today even the smallest impurities – the micro pollutants – can be detected. Later this century this will be done automatically. The Royal Society of Chemistry's report, *Sustainable Water: Chemical Science Priorities,* envisages a network of sensors at key locations capable of operating independently and testing for dissolved oxygen, conductivity, pH, nutrients, pesticides and chlorinated hydrocarbons. The data would be communicated by wireless technology to computers for statistical processing and interpretation by analytical chemists.

An example of what might be done to monitor water is assessing the amount of ammonia it contains. This unpleasant component is removed by special bacteria which can oxidise it to nitrite and nitrate. The danger is that it could be converted to nitrous oxide (N_2O) which is a highly potent greenhouse gas. A sudden rise in the amount of this indicates a breakdown in wastewater treatment process. At Cranfield University a spin off company, Water Innovate, has devised an automatic monitor which they call N-Tox and which measures the amount of nitrous oxide in the air above wastewater. It pumps air through an infrared detector linked to an automatic sensor and data storage device.

Another example of remote monitoring is in Minnesota where there is a water source at risk of being contaminated during flooding from land. The water source is continually analysed for temperature, pH, turbidity, dissolved oxygen concentrations, nitrate and phosphate levels, with the data being transmitted back to the laboratory run by Paige Novak and her colleagues in the Civil Engineering Department of the University of Minnesota.

Over the years there have been some dramatic cases of unsuspected metal pollutants in the water supply. Perhaps the most famous case was that featured in the 2000 film *Erin Brockovich,* starring Julia Roberts, which is the story of a feisty young woman's campaign to clean up a water supply that was contaminated with **chromium**(VI) which is a known carcinogen. She believed this was coming from the Pacific Gas and Electric Company who ended up paying $333 million to settle a class action lawsuit. It now turns out that they may not have been guilty because chromium(VI) ions are still present in the water.[9] At least that's

[9] These could be formed naturally by the action of iron and manganese ions, which are present in the local aquifer, reacting with other chromium ions to produce the chromium(VI) ions.

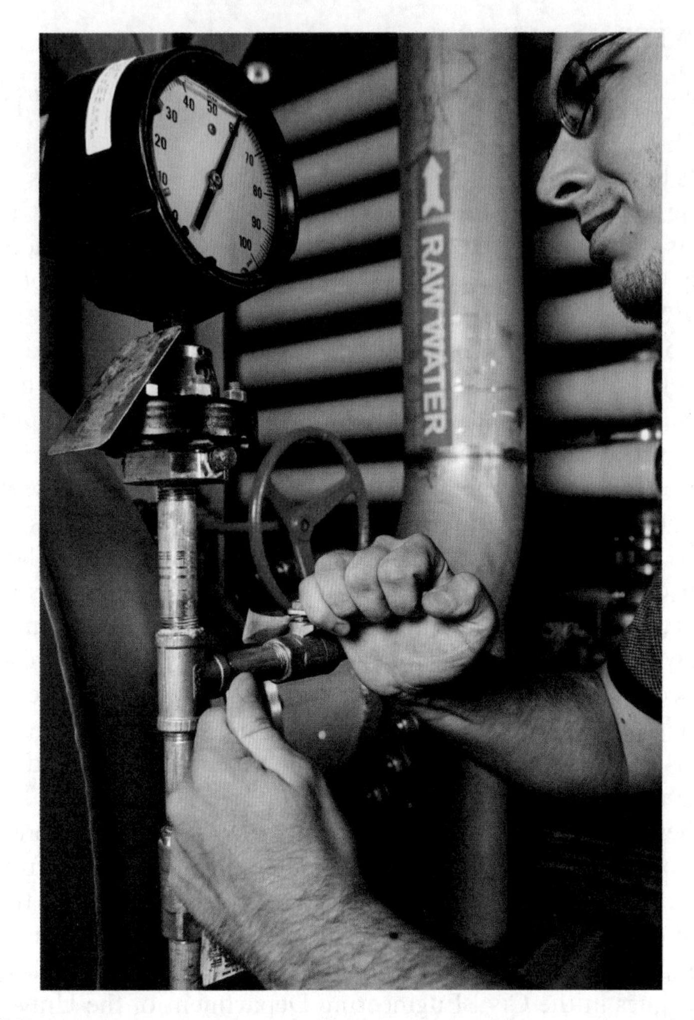

Figure 2.2 Chemists can make even the foulest water clean enough to drink.

what Ana Gonzalez of the University of California suggested in work published in 2005.

A larger problem of contaminated drinking water was that faced by the villagers in West Bengal, India and Bangladesh. Their wells were installed in a drive by the United Nation's Children's Fund (UNICEF) to provide safe drinking water for a population that had traditionally taken its water from contaminated streams, rivers and ponds, and indeed there was a decrease in illnesses from water-borne diseases, especially among children. Analysis of their water showed that there were no heavy metals present. However, the analysts had not tested for arsenic because this was not suspected to be a contaminant.

The first cases of chronic arsenic poisoning were identified in 1983. Those affected suffer from hyperkeratosis, which is characterised by a thickening of the skin on the palms of hands and the soles of feet which can lead to cracking and infections. Tests soon showed that drinking water was to blame, and in Bangladesh more than 20 000 wells were found to have arsenic levels in excess of 100 ppb and some had ten times this level. The World Health Organisation says that arsenic in drinking water should not exceed 10 ppb, although water with 50 ppb is considered to have no observable effect and the US accepts this higher level. A quick method for testing whether water is safe to drink has been developed by Jan Roelof van der Meer of the Swiss Federal Institute for Environmental Science and Technology. He has modified strains of *E. coli* bacteria so that they emit a green fluorescence when they come into contact with water that has more than 4 ppb arsenic.

Removing arsenic is not difficult. It simply requires it to be passed over alumina (aluminium oxide) which absorbs it and forms an insoluble aluminium arsenic salt. A cheaper method uses rusty iron filings, which are excellent at removing arsenic by forming insoluble iron arsenite. A simple filter consisting of layers of sand, charcoal, and iron filings, costs very little, is long lasting, and will purify arsenic-containing well water very efficiently and more than 20 000 such units are now being used in Bangladesh.

Never has it been more important to analyse water supplies. There are fears that terrorist groups might deliberately target them with nerve gases or biological agents such as ricin. This toxin can be extracted from castor oil beans and it is deadly in the tiniest of doses. The symptoms of ricin poisoning will generally be mistaken for other diseases and proper diagnosis can be delayed for several days, by which time it will be too late to save the victim. Suspect samples of ricin-contaminated materials can be detected and positively identified using enzymes, and there are simple testing kits to do this. A typical device is the AbraTox kit produced by the Advanced Monitoring Systems Center which uses the bacteria *Vibrio fischer* which emits light. When it encounters toxins in a sample of water this light is diminished in direct proportion to the amount of toxin, and it can detect as little as 15 ppm of ricin. (It will also detect nicotine, Botulinum toxin, cyanide and nerve gases.)

Another detection method relies on strips that change colour. A sample of suspected water is applied to one end of the strip, and as it diffuses along the strip it passes several stripes which have been impregnated with specific antibodies that have been labelled with coloured dyes. If they change colour they not only show that a toxin is present but identify which one it is.

2.4 WASTEWATER TREATMENT

Industrial wastewater and even sewage can be cleaned up and reused.

In some parts of the world vast amounts of urban wastewater are simply discharged into rivers or the sea. In Asia two thirds ends up being disposed of this way, and it is a wasted resource because it is possible to clean it up and recycle it. It is even possible to clean up foul water so that it is as fresh as tap water. Advanced water purification is already a reality in parts of the US, and in the Santa Ana region of California it provides more than 250 million litres per day. They have the largest water reclamation plant in the US and it removes organic chemicals, pathogens and pharmaceutical residues, producing water that is purer than tap water.

The Santa Ana reclamation process involves many stages. First the sewage is filtered to remove solid items, and then allowed to stand so that gritty materials like sand and coffee grounds can settle out and be removed. Next it is treated with a flocculant and that too is allowed to settle out. The water is then aerated to encourage bacteria to go to work breaking down faecal matter, followed by a third settling stage. The water is now ready for advanced treatment, which includes adding **sodium hypochlorite** to disinfect it, followed by micro filtration through polypropylene membranes with pores of around 2 microns to remove things like dead bacteria.

Finally there is reverse osmosis – see below – after which hydrogen peroxide is added to the water. When this is exposed to UV light it forms hydroxyl radicals and these react with any residual traces of impurities. Such water is now fit to drink, not that it is so used. It goes to a large pond where it slowly percolates through the ground and into the water table, a process that takes six months as shown by adding **noble gas** traces to it. Only then can it be extracted *via* wells and returned to the mains supply.

There are micro pollutants in wastewater which previously were unsuspected, and which can survive conventional waste treatment processes which remove the macro pollutants. The micro pollutants include several drugs, such as painkillers, antibiotics, antidepressants, contraceptives and beta-blockers, and chemicals like the insect repellent *N,N*-diethyltoluaminde (DEET) and the antiseptic **triclosan**. Existing water treatment does not completely remove these and they can be detected in rivers and even ground water.

Hypochlorite treatment will destroy many chemicals but ozonation is more effective, and if this is followed by filtering through activated carbon then all micro pollutants are removed. Another method of destroying them has been suggested and that uses ultrasound in the range 20 kHz to 10 MHz. The ultrasonic resonator's high-frequency sound waves produce minute bubble-like cavities in the water into which molecules diffuse, there to be destroyed when the cavity collapses with a violence that generates a flash of high temperature.

Micro-filtration membranes will remove microbes and nano-filtration membranes can even remove calcium and magnesium ions very effectively, but membrane technology is expensive because it requires high pressures.

Some sewage treatment works are required to remove both nitrate and phosphate if these pose a threat to bodies of water, since nitrate and phosphate act as nutrients for algae which can then grow to such an extent that they threaten other life forms. Nitrate can be removed by denitrification processes which are biologically based. *Nitrospira* species use nitrate as a source of oxygen, releasing the nitrogen as a gas to the atmosphere. As we will see in Chapter 3, phosphate can also be removed by bacteria which accumulate it and these can be harvested and even used as fertiliser. However, most phosphate is removed by precipitating it as either aluminium or iron phosphate because these processes are easier to operate and more controllable.

The easiest and most commonly used method of reducing pollutants in wastewater is to allow it to trickle through a bed of activated carbon which will absorb almost all of the unwanted substances. When the carbon has become saturated with pollutants then it can be regenerated, although this requires high temperatures and as much as 20% of the carbon is lost. Alternatively, it can be disposed of by incineration or dumped in a landfill site.

A much more efficient method of regenerating water, albeit on a smaller scale, is to expose it to oxidative electrolysis and this is what the Arvia process does. This has been devised by Chemical Engineers at the University of Manchester. It can operate continuously and works by quickly absorbing pollutants on to special carbon particles, which are then carried by a flow of air bubbles to an inner chamber while the cleaned-up water passes out of the top of the chamber. Meanwhile, the particles settle and are exposed to a direct current of electricity, which has the power to oxidise the pollutants and destroy organic matter, pesticides, chlorinated organics, oils and solvents within a few minutes. The regenerated particles are then drawn out of the bottom of the inner chamber by the flow of incoming effluent and they go through the process again. All this takes place at normal temperatures and it promises to enable the same water to be recycled and reused within a manufacturing company. So far the researchers have shown that a simple 13 amp supply can treat as 200 cubic metres (200 000 litres) of water per day.

The chemical industry uses a lot of water but this is declining in Europe as companies concentrate on reclaiming water, while others re-use water that has passed through municipal sewerage plants. This is done at one site in Terneuzen, the Netherlands, where it meets a quarter of the site's needs along with a similar amount that is recycled within the plant. However, the wastewater from chemical and pharmaceutical laboratories has to be specially disposed of, sometimes *via* an incinerator, if it is suspected of containing biologically active molecules or new chemicals whose toxicity is not known.

2.5 WATER FOR IRRIGATION

The agrochemical industry is finding ways for farmers to save water.

In global terms the biggest consumer of fresh water is agriculture, and in dry regions of the Earth this may reach 90%. Reducing this demand will be met in three ways: first, by planting hardier GM crops which need less water; second, by preventing rainwater run-off; and third, by helping crops to manage water shortages better.

Figure 2.3 In many countries of the world, almost all the available water is used to irrigate crops.

The first of these methods will reduce the need to irrigate by modifying the crop itself so it becomes more drought tolerant. This has been happening in the corn belt of the USA where vast areas of land are used to grow maize (corn). Dryland maize has been genetically modified to include drought-tolerant genes from wild maize. It doesn't need irrigating and can yield around 5–7 tonnes per hectare. Re-introducing more native traits could permit other ways of combating water stress to be built in, and protect plants at particularly vulnerable stages in their development, such as when they are flowering. Roots are the key to survival, so genes that improve root growth and root depth might be incorporated. Water loss from leaves might also be cut by reducing the number of stomates, from which water escapes.

China has 7% of the world's arable land but has to feed more than 20% of the world's population, and the area under cultivation is decreasing due to rapid urbanisation. Rainwater run-off is a major problem in hilly areas, and this happens when the soil is disturbed. No-tillage farming can increase food output while reducing the demand for water. Tillage is any agricultural practice which disturbs the soil, such as ploughing and pulling up weeds. The alternative is to kill weeds with a non-persistent herbicide, such as **paraquat** (aka Gramoxone) or glyphosate (aka Round-up), and this cuts water loss significantly.

In the mountainous areas of the Yangtze River, there are peach orchards and tea plantations where water run-off and soil erosion was a problem. Test sites were marked out with retaining walls which directed water run-off to a tank, the contents of which were analysed. The results showed that water loss from sloping land that was weeded by hand was twice that on land treated with paraquat. Tilled land lost 420 m^3 of water per hectare per year, whereas the loss from paraquat-treated land was 220 m^3. Loss of topsoil along with the rain was correspondingly much greater on land that was tilled. Topsoil loss was 2.3 tonnes per hectare under normal weeding, compared with only 1.3 tonnes from land treated with weedkiller.

Paraquat has been helping farmers around the world since the 1960s and is now used by millions of growers. It is extremely safe. It leaves no residues on the crop, it leaves the roots of weeds in the soil which contributes to soil conservation, and it replaces the back-breaking toil of weeding by hand. Paraquat only affects the green parts of the plants on which it falls, and because it has no effect on trees it can be used to weed areas of olive trees, coffee bushes and grapevines. It is fast-acting, not washed off by rain and the paraquat is deactivated on contact with soil so it presents no threat to the next crop to be planted.

The third method of getting plants to cope with water shortage is to persuade them that it's not happening, and that can be done by blocking an enzyme in the plant. When a plant is under stress from lack of water it produces a simple hormone: ethylene, which tells it to stop growing or to come to fruition more rapidly, both of which reduce crop yields. The ethylene enters special receptors on plant cell membranes and triggers them to act.

Blocking the ethylene-sensitive enzyme can boost crop yields by 10%. The chemical to do this is known as 1-methylcyclopropene, (aka 1-MCP, trade name Invinsa). Research in greenhouses demonstrated very well that 1-MCP made plants grow taller and produce more of their crop. However, this molecule is a gas, which poses a problem when it comes to using it in the field. Chemists have found a way of surmounting that problem. The trick is to trap the gas inside α-cyclodextrin molecules. These water-soluble carbohydrates are easily made from starch and they have a band-like structure which is able to wrap around smaller molecules. In this way the 1-MCP can be delivered to where it is needed.

Before we leave the topic of water for agriculture, I should point out that as the world moves over to biofuels then the demand for water is bound to increase. Water and energy are closely linked – or will be when the world relies mainly on biofuels, which are the subject of Chapter 4. A litre of biofuel requires a thousand times more water for its production than a litre of petrol produced from oil. In terms of litres per km travelled, all the alternatives to oil require more water. For example, to produce the energy to power an electric vehicle for one km requires 23 litres of water, whereas for biofuel this is well in excess of 100 litres.

2.6 SEAWATER

There's plenty of water in the sea and chemists can make it drinkable.

'Water, water everywhere, nor any drop to drink' so bemoaned the sailor in Samuel Taylor Coleridge's poem of 1798, *The Rime of the Ancient Mariner*. So will many in the future bemoan their plight unless chemists make desalinated seawater cheap and plentiful, and this they are doing.

When you have a semi-permeable membrane with pure water at one side and salt water at the other, then pure water will diffuse through the membrane in order to dilute the salt water. That's osmosis. As the height of the salt water rises its pressure increases and eventually to such an extent that diffusion stops. However, if you apply a high enough pressure to the salt water then you can reverse the process and squeeze out pure water from the salt solution. That's reverse osmosis. Given the right type of polymer membrane and a high enough pressure, seawater can be desalinated and most desalination currently requires pressures of around 70 bar (approximately 70 times atmospheric pressure). Pressures as low as 10 bar might be possible when the newer plants come on line with better membranes.

One day it may be that all kinds of communities, from coastal cities to luxury liners, will be provided with cheap, clean water. Some are already using desalination as their source. The island of Lanzarote in The Canaries began operating a desalination plant in 1964 and its 120 000 residents are supplied from it. It also has to provide water for more than a million tourists per year. In the 1970s they changed over to reverse osmosis and while the early plants used about 12 kWh of electricity to produce a cubic metre of water, the newer processes have reduced this to less than 4 kWh. Even this was improved in 1997 by installing a pressure-exchanger which makes use of the high pressure of the outgoing water to pressurise the incoming water. Now energy requirements are down to around 2.5 kWh per cubic metre of water. On the Canary island of El Hierro a new desalination plant will soon operate using only renewable energy from solar panels and wind generators which will supply all of the 10 000 inhabitants with fresh water.

The Australian city of Perth gets 250 000 cubic metres (m^3) of water a day from a desalination plant which provides a sixth of its needs, albeit at a cost of around $250 000 per year. Singapore has a more energy-efficient desalination plant which provides 135 000 m^3 a day but costing only $65 000. There are companies who claim that even this rate could be halved so that a cubic metre (1000 litres) of water could be produced for as little as 25 US cents. Desalination systems are now a billion dollar a year industry and growing at an annual rate of 7 per cent. It is not only seawater which can be purified this way.

Early desalination plants relied on distillation to produce drinkable water and a lot is still produced this way. They require seawater to be heated and the vapour rising from it to be condensed and collected. The more concentrated seawater is then piped back into the sea. Heat exchangers can make the process more efficient. The passing of the seawater to successive chambers – generally six in all – of lower and lower pressure ensures that almost all the heat is used. (The heat might usefully come from a steam turbine generating plant.) Such plants were built in the 1950s and 1960s and as they became larger they could eventually produce as much as 50 000 m^3 per day. The biggest bugbear is the formation of scale in the plant which reduces its efficiency, although scale inhibitors can be added to the water to prevent this happening.

Desalination relies very heavily on a copper nickel alloy (90% copper and 10% nickel) because this resists bio-fouling, while being stress and corrosion resistant. The membranes for desalination are made from various polymers, namely **cellulose acetate**, polyamide, and polysulfone.[10] Reverse osmosis requires the water being treated to be free not only of suspended solids, but also bacteria or anything that might disrupt the system. Over time the surface of desalination membranes get blocked with microorganisms such as algae. When this happens, the pressure has to be increased but eventually the membranes do have to be cleaned. This contamination can be prevented if chlorine is added to the water. However, the chlorine has then to be removed before desalination, because it attacks the commonly used polyamide membranes.

Researchers in the US and Korea have come up with a membrane made of polysulfone which is not affected by chlorine. The membrane consists of a micron-thin layer of porous polysulfone supported on a thicker layer of polyester. Newer nano composite membranes now being developed have nano particles of zeolites embedded in the polymers. These attract water and they also act as pores in the membrane. They are being produced by the company NanoH_2O.

Other more revolutionary membranes are also on the horizon. A group headed by physicist Olgica Bakajin and chemist Aleksandr Noy, based at the Lawrence Livermore National Laboratory (LLNL) in California, have developed membranes with carbon nanotubes pores and there are 250 billion of them per square cm. What is remarkable about the new membranes is the speed with which water molecules can flow through them, more than 1000 times faster than expected. Their new devices might one day be important as a method for desalinating seawater.

Desalination plants are now in operation in 120 countries, producing more than 35 million m^3 per day, and this will double by 2015 and it will cost $95 billion per year. Today there are 15 000 desalination plants in action and they are not only to be found in countries where natural rainfall is low, such as the United Arab Emirates, Kuwait, Southern Spain, North Africa and The Canaries, but in the US and Australia. California plans to build 20 desalination plants, the largest of which would produce 400 000 m^3 of water per day. In China there are plans to produce 2.5 million m^3 of desalinated water by 2015. Even in the UK, Thames Water is building a desalination plant at Beckton in east

[10] The main membrane suppliers for reverse osmosis are the US companies Dow and Hydranautics, and the Japanese firms Toray and Toyobo.

London which will be able to provide up to 140 000 m^3 of water per day during periods of low rainfall, or when one of their other treatment facilities is out of action.

2.7 EXTREME WATER

Water can be made do almost magical things.

Finally in this chapter, we look at the new ways in which chemists use water and with which they can carry out some remarkable processes.

There are three kinds of water which scientists use for their research and which certain industries also use: high-temperature water, supercritical water and ultrapure water. All have rather special applications.

Water's chief role is to be a solvent, but H_2O shouldn't be a liquid at all – it should be a gas like its smelly sister hydrogen sulfide (H_2S). What gives water its ability to be a liquid are its hydrogen bonds which allow its molecules to stick together. When hydrogens are bonded to oxygen they become slightly positively charged and this makes them attracted to the oxygen atoms of other molecules, which are slightly negatively charged. The result is a chemical which melts at 0 °C and boils at 100 °C. (Hydrogen sulfide, H_2S, melts at *minus* 82 °C and boils at *minus* 60 °C.)

Like dissolves like. Oil and water don't mix. While these common rules of thumb may be true with ordinary water, they may not always apply. Because water molecules have positive and negative ends they attract other species that are charged, making water an ideal solvent for salts like sodium chloride which consists of the ions Na^+ and Cl^-. However, water is not so good at dissolving molecules which are not polar, such as hydrocarbons, and this is why oil and water don't mix and why water is not a particularly good solvent for lots of organic chemicals which tend also not to be very polar. Organic solvents have then to be used, such as toluene and acetone. However, chemists are now trying to avoid such solvents because they are volatile and add to the unwanted greenhouse gases of the atmosphere.[11]

Water may be a remarkable chemical but it can be made to do even more remarkable things. High-temperature water is not only a good solvent for ions but for organic molecules. This is water which has been heated above its normal boiling point of 100 °C and is kept as a liquid by applying pressures; the higher the temperature, the higher the pressure

[11] Water in the atmosphere acts as *the* greenhouse gas, keeping this planet's temperature in a range within which life can flourish. Were the Earth as dry as the Moon, then its average temperature would be below minus 20 °C.

needed to keep it a liquid. As the temperature goes up so the density of the water goes down, and in the usual working range of high-temperature water, which is 250–300 °C, the density is 70% of normal water. The polarity of the high-temperature water also decreases, so that all kinds of organic molecules will now dissolve in it and it becomes a good solvent in which to perform chemical reactions. Not only can oxidation reactions be carried out in high-temperature water, but so can reactions which form carbon–carbon bonds, as well as hydrogen/**deuterium** exchange reactions. As yet it is relatively little used because of the capital costs involved, but it seems likely that in future it will be employed.

Supercritical water is even hotter than high-temperature water. The critical point of water is 374 °C, above which it loses its ability to be a liquid no matter how high the pressure. Then we have supercritical water. The critical pressure at this temperature is 218 atmospheres (220 bar). Supercritical water has even been found to occur naturally. In 2008 it was discovered being emitted by two vents (black smokers) situated 3 kilometres down on the sea floor of the Atlantic Ocean just south of the equator. Its temperature was 407 °C and the pressure was 300 bars. Supercritical water is a curious fluid that will dissolve almost anything and is used to destroy chemical warfare agents, such as nerve gases.

Ultrapure water is needed in analytical and research laboratories, and in pharmaceutical and semiconductor manufacture. (Supercritical water has also to be ultrapure.) Producing a silicon wafer with its 100 semiconductor chips requires almost 10 000 litres of ultrapure water.

Drinking water is fine if it is 99.9% pure, but this is not good enough for researchers and pharmaceutical companies which need water to be at least 99.995% pure. Such water is attainable by **ion exchange** and reverse osmosis. However, sometimes only 99.999 999 999% pure water will suffice. This is made by putting 99.995% water through yet more ion-exchange and membrane processes, combined with intense UV treatment to destroy any organic material that might be present. Units are available in laboratories which can produce litres of ultrapure water per hour. Ultrapure water cannot be stored in glass or plastic containers because these will contaminate it by releasing material from their surface, even though that release is tiny. Instead it has to be stored in vessels made of ultrapure tin.

The pharmaceutical industry requires ultrapure water and it has been at the forefront of developing methods of producing this using reverse osmosis, ion exchange and micro-filtration. Now there is the micro-bioreactor (MBR) system which combines filtration, followed by biological decomposition, and finally hollow-fibre filtration. This last technique involves the use of membranes made up of hollow, spaghetti-like fibres made from hydrophobic polymers, such as polyvinylidene fluoride, which contain billions of microscopic pores. Apply gentle suction with a vacuum pump, and only water and nothing else can be sucked through the membranes.

Those then are seven important ways in which our use of water involves chemists and chemistry. Some countries like China are investing heavily in this area to bring clean, fresh water to more of their population, and they are installing the most advanced technologies available, including state-of-the-art desalination plants. For the developed world, the issue is one of reducing 'virtual' water usage, and forgoing the doubtful benefits of bottled water. There are scores of ways that water can be saved and in the end it is up to individuals to ensure that they waste as little as possible. Fresh water should be everyone's right and must become a sustainable resource, no matter how many people there are in the world. Chemists can make it so, given the opportunity.

CHAPTER 3

Health and Chemistry

[A word in **bold** means there is more information in the Glossary.]

Healing drugs and chemistry go hand in hand and many of the ills that flesh is heir to can be treated by using chemical products to control the errant reactions within the body. Taking the right pill can relieve many illnesses. Seven different health problems and their respective drugs are discussed in this chapter: insomnia, obesity, flu, multiple sclerosis, infections, cancer and asthma.

Living to a hundred was a rare event back in 1900, when there were only about a hundred such individuals in the UK, whereas today there are more than 10 000 centenarians. According to the UK National Statistics the explanation for this increase is better medical treatment, hygiene, sanitation, housing and nutrition. Of course the most important factor is gender, because 90% of these centenarians are female. However, many of them would not have reached this venerable age without the pills and potions discovered by chemists in the pharmaceutical industry.

Probably the most important step in reducing the incidence of human disease was a supply of clean water and underground sewers – and keeping these two systems apart. Personal hygiene is another way of reducing the risk of ill-health, and chemistry comes into the picture in the shape of products with which to wash our hands and clean kitchens and toilets. Finally we need to keep our bodies in good shape if we want to live as long as our genes will let us. There are seven things we can do to ensure this:

1. Stop smoking.
2. Maintain the right weight or **body mass index (BMI)**.

A Healthy, Wealthy, Sustainable World
By John Emsley

Published by the Royal Society of Chemistry, www.rsc.org

3. Take more exercise.
4. Limit our intake of alcohol.
5. Eat five portions of fruit and vegetables a day (and not including potatoes).
6. Drink at least a litre of (tap) water a day.
7. Increase our intake of fibre.

There is not much more that we can do to maintain our health – although a good night's sleep also helps. When we fall ill then we should seek help from a doctor, although we may be ensnared by alternative medicine, thanks to its high profile in the media. What they offer has no scientific basis and are questionable treatments at questionable costs.

Common Sense 8: Homeopathy can cure many ills

Wrong. Homeopathy relies entirely on the placebo effect because generally all you are given is water. Homeopathy appears to work on a third of those who take its medicines, which is the same proportion of people who report a benefit when they take part in **double-blind** drug trials, but who are given the placebo disguised as the real pill. The same placebo benefit can occur with all alternative treatments. If the belief is strong enough, it really can be beneficial, especially if the illness is more psychological than real. Also, many medical conditions are self-limiting and will cure themselves given time, which is why taking homeopathic water or a placebo may appear to affect a cure.

In this chapter, we will look mainly at the kinds of treatments which scientifically trained doctors and specialists prescribe, but first we will look at two ailments which we can diagnose ourselves and for which we can purchase over-the-counter remedies. These conditions are lack of sleep and being overweight. Both seem to have become endemic in developed societies.

3.1 SLEEPING PILLS

Sleep like a baby! Well now you can – at least for a while.

Demand for sleeping pills continues to grow and around the world expenditure on them exceeds $10 billion a year.

Figure 3.1 Sleep is all about brain chemistry, and chemistry can now ensure a good night's rest.

We would all like to go to bed when it suits us, fall asleep rapidly, and stay so for the 7 or 8 hours that should leave us refreshed and ready for a new day. However, for many of us a full night's sleep is something we can only dream about. People today are less physically tired, but are more mentally tired and that's not conducive to a good night's sleep. According to the US Department of Health, every year more than half of the adults in the US suffer from insomnia at some point, and women suffer more than men. Insomnia can be due to many causes, such as illness, work stress, family worries, recreational drugs, jet lag, or shift work. Whatever the reason, you can turn to pills for help.

The traditional sleeping draught is of course the chemical ethanol (alcohol) and this is probably the most commonly used pre-bed relaxant. The first kind of properly devised sleeping pills were those based on **barbituric acid**. In itself this does not cause sleepiness but its many derivatives do, and one of these, barbital, was launched as Veronal in 1903. It had to be a prescription-only drug because it was potentially dangerous, the reason being that the effective dose was near the toxic dose. Some celebrity deaths were attributed to overdoses of Veronal, including the film star Judy Garland who died in 1969 aged 47.

In the 1960s the benzodiazepines appeared and there were many variants of these, of which temazepam was the best known. They were better than barbiturates, although they had side effects associated with long-term use, in that patients could become addicted to them. Benzodiazepines act on a brain chemical called gamma-aminobutyric acid (aka GABA) and they induce sleep by controlling this.

Non-benzodiazepines medicines have replaced the benzodiazepines in recent years. They act on a receptor in the brain known as omega-1 and this induces sleep. Some other drugs have sleep as a side effect, such as antidepressants and antihistamines. There appear to be several sleep centres in the brain and more than one kind of molecule will trigger them to act. What we want from chemistry is a medicine that will ensure that we sleep as naturally as possible and which over-rules the worries that prevent us from getting it. Doctors now have several kinds of pill to choose from. The non-benzodiazepines are zolpidem (aka Stilnoct, Ambien, *etc.*), zaleplon (aka Sonata), zopiclone (aka Zimovane), eszopiclone (aka Lunesta) and indiplon. The last of these has yet to be approved but is said to be one of the most potent. These drugs have short half-lives in the body – Sonata's is only 30 minutes – so they can even be taken any time insomnia occurs, even if that is the middle of the night.

Two particularly successful sleeping pills are zolpidem which was introduced in 1993 and eszopiclone which was launched in 2005. Together sales of these two drugs amounted to $3 billion in the US in 2006. What zolpidem may also do is help repair the brain and has been known to *wake* people up who have been in a coma for weeks and months, and this aspect is being tested at the current time. Coma victims were given it on the assumption that if they actually went to sleep they might then wake up. There are aspects of zolpidem which mean it will always be a prescription-only drug, because it is dangerous to drive while under its influence and it has been known to cause sleepwalking. There are possibilities of its being misused as a recreational drug which produces visual effects and hallucinations. Zolpidem can only be a temporary treatment because it loses its effectiveness after two weeks.

Eszopiclone is also a prescription drug that acts on certain receptors in the brain. It works quickly and only 2 or 3 mg are required, and even as little as 1 mg for older people. There are side effects, such as feeling dizzy, but these are likely to be experienced only by those who stand up shortly after taking the drug. Nor is it a drug to be taken by people who are prone to depression.

Actelion, a Swiss pharmaceutical company, has discovered another kind of sleeping pill: almorexant. This acts on the orexin system which was only discovered in the late 1990s and which releases a hormone that keeps us awake; block this and sleep should come. Orexin-RA is currently undergoing tests and it could be on the market by 2012. It increases the dream phase known as REM sleep (short for rapid eye movement) which is the period when the brain is hardwiring memories of the previous day. REM periods account for around a quarter of our time asleep. The lack of orexin is the cause of the sleep disorder known as narcolepsy. Samples of cerebrospinal fluid taken from narcoleptics showed that they contained almost no orexin, and autopsies of the brains of those with this condition and who had died were found to have withered orexin neurons. Those suffering this disorder can fall asleep at any moment and there are now drugs to treat this condition, such as modafinil and adrafinil. These have the potential for keeping people awake, and are taken by aircrews on long flights.

However, if we are not sleeping well, and we know the cause is temporary, then we can turn to over-the-counter treatments and there are some quite effective ones. The best known brand of sleeping pill is Nytol (Benadryl in the US) whose active ingredient is the antihistamine, diphenhydramine hydrochloride. This chemical easily passes the blood–brain barrier and there it encourages you to fall asleep and then will keep you asleep. It is meant to be used for a short while, say two weeks or so, when you know what is causing your sleep problems. The dose is 50 mg and provided you are not young or drinking alcohol then there is little risk of side effects.

Another way to induce sleep is to take some of the chemical which the brain produces when it wants to go to sleep, **melatonin**, which traditionally has been the drug you took to counteract jet lag. The peanut-sized pineal gland at the centre of the brain produces this hormone in response to nightfall, or at least in response to the amount of light entering the eye and the level of melatonin increases in the early hours of night and then begins to fall as we approach dawn. As we reach old age, our ability to produce melatonin decreases. A 3 mg dose of melatonin is enough to send a person off to sleep within a few minutes. Melatonin was heavily promoted in the 1990s, with some even referring to it as the body's own wonder drug saying it could cure or prevent all kinds of illnesses and even cancer. There has been no convincing support for this but a lot of research has gone into understanding the way it acts.

Melatonin has several roles; in humans it helps us to adjust our sleep patterns to the daily rotation of the planet and its annual cycle round the sun. It also controls our body temperature, reducing it slightly during the hours of sleep. In sheep and deer, melatonin signals the breeding season, while in other animals it causes moulting, and for some it determines the time to hibernate. Ramelteon (aka Rozerem) is based on melatonin and was approved in 2006 as the first of a new class of prescription sleeping pill.

Another chemical which can induce sleep is oleamide which is a derivative of oleic acid. This occurs naturally in animals, including humans, and was observed to build up in the cerebrospinal fluid of cats which had been deprived of sleep for 22 hours. When a 5 mg dose of this molecule was injected into other cats they quickly fell into a deep sleep, and while they rested the molecule was removed by an enzyme called *oleamide hydrolase*. Tinkering with the oleamide molecule, by shortening or lengthening its chain with an extra carbon, also produced other substances which were effective but as yet none has emerged as a sleeping aid for humans.

3.2 OBESITY

This modern affliction is spreading. Can chemistry help prevent it?

Obesity is causing concern in developed countries where food is plentiful and most occupations involve very little in the way of physical exertion. In the UK around 60% of the population are already overweight and a third of them are classed as obese.

There are lots of products which claim to help reduce weight and some are based on various natural extracts, such as caffeine which is said to burn off fat. Their promoters talk of them being ‘fat-binders’, ‘carb-blockers’, and ‘appetite suppressants’, and promise not only that you will lose weight, but will have more vigour and sexual desire. Of course with so many confounding variables involved in dieting, there is no way that any of these claims can be proved scientifically. However, some products might work, such as those containing hoodia, and at least there is some scientific backing for this. Extracted from the Southern Africa cactus *Hoodia gordonii,* this product is reputed to suppress appetite, and tests with rats have shown that the carbohydrate which it contains, known as P57, can affect the brain in a way that suppresses the desire to eat. Rats fed hoodia reduced their food intake by around 50% in the next 24 hours.

Weight-reducing drugs have been around since the 1930s. The first was 2,4-dinitrophenol (aka DNP). This had been used to make munitions in World War I and the women workers who came into contact with it reported losing weight. When it was prescribed as an obesity drug, it clearly was effective but sometimes it led to high fevers and occasional fatalities. Its replacement was amphetamine which was prescribed to treat depression, but it was noticed that while those on this drug also lost weight some patients had raised blood pressure and mental disorders. It too was discontinued as an obesity treatment.

In 1973 fenfluramine (aka Pondimin) began to be prescribed to treat obesity. It worked by boosting the level of serotonin in the brain which not only reined back our desire to eat, but made people feel better about themselves. What was not apparent was the effect it could have on the heart where it caused a thickening of tendons. A study of 5700 users of the drug showed that 15% of them were affected this way, and consequently the drug was withdrawn from the market in 1997.

Another obesity drug, rimonabant, was withdrawn in 2008 when the European Medicines Agency said that its benefit as an anti-obesity drug was outweighed by an increased risk of its causing mental disturbances.

Today more than ever, there is a market for safe anti-obesity drugs. Of course we can bring our diet under control without drugs simply by eating sensibly. We can take advantage of what food chemists have achieved in making 'diet' and 'light' alternatives by substituting complex carbohydrates for fats and artificial sweeteners for sugar.[1] However, what some people want is not the discipline of a 1500 Calorie[2] diet, a cure that takes weeks, but a pill that will enable them to continue to eat what they want without putting on weight. What chemists can offer is nothing like that, nor should we even try to produce such a drug. What has been produced is a drug that will shorten the time it takes to lose weight whilst on a diet.

Orlistat acts in the body in a way that prevents fat in the diet from being absorbed. It does this by binding to the enzymes which are needed to digest the fat. In February 2007 the FDA finally approved orlistat as an over-the-counter drug and in January 2009 the EU also approved it. You can buy it at a pharmacy under the brand name of alli (*sic*) at around £40 for a two-week supply. Of course, being a product of the pharmaceutical industry it has attracted scare stories of nasty side effects, most notably oily diarrhoea. It may increase the risk of kidney stones and those with a history of these will be advised by the pharmacist not to take it. Early indications suggested a higher risk of breast cancer, but it later transpired that the three women who developed the cancer already had signs of the disease before the trial started. A two-year study published in 1999 found that there was no higher risk of breast cancer among women taking orlistat.

Orlistat is produced by GlaxoSmithKline. (There is also a stronger dose available by prescription as the drug Xenical and produced by Roche.) Orlistat cannot be sold to people under 18 or to someone with a body mass index below 28. The leaflet which comes with alli promises

[1] This assumes that the result will be fewer calories, but not if a lot more carbohydrate is used in place of the fat.

[2] A food Calorie is 1000 ordinary calories, aka a kilocalorie.

users that if they stick to a 1400 Calorie diet, with a fat intake of 15 grams,[3] and take alli, then for every two pounds they lose by dieting they will also lose a pound by taking the drug.

Orlistat has a specific and long-lasting effect in the gut and it works by reacting chemically with the active site of the *gastric lipase* enzymes which are there to digest fats and oils. They do this by releasing the fatty acids from the glycerol molecule to which they are attached, thereby allowing them to be absorbed into the body. A 60 mg dose of orlistat taken three times a day before meals will ensure that about a quarter of the fat in the diet is not broken down, and of course the meal being eaten should be low in fat in any case if the eater is intending to lose weight. (If a doctor prescribes orlistat, then the dose is much larger and will ensure even less of the fat in the diet will be absorbed.) What has to be guarded against is interfering with the intake of the fat-soluble vitamins, A, D, E and K, so a supplement of these should be taken at bedtime. The loss of weight begins within a week or so. Tests on orlistat over a period of a year showed that about half those on the drug reduce their weight by 5% or more, and a quarter achieved a loss of around 10%. What was also noticeable was a decrease in type 2 diabetes.

Chemists are still seeking an answer to obesity and one which works quicker than orlistat. Such a drug was announced in 2004, a peptide-based drug which targets fatty tissue directly and breaks this down so that the body has to reabsorb the fat and use it. At least that's what happened with mice which were fed until they were obese and then injected with the drug. After a month they had lost 30% of their body weight and returned to their normal size. Whether such a treatment will become available to doctors treating obesity remains to be seen. In any case it will be several years before it successfully completes a testing schedule – it if ever does.

3.3 FLU

If a deadly pandemic of flu develops, then antiviral drugs will be needed.

A virus carries out its work in four stages. First it attaches itself to a living cell. Then it releases into the cell the components needed to replicate itself and these are genes and enzymes. Then it takes over the host cell's replicating mechanism to make copies of itself. Finally it releases itself from that cell and goes off to infect other cells. Knowing this, it is possible to seek drugs which will interfere with each stage.

[3] 15 grams of fat provides 135 Calories.

Flu is caused by the influenza virus of which there are many mutations, and occasionally one emerges that is particularly dangerous, such as Spanish flu which appeared near the end of World War I. It was called this because only the Spanish authorities admitted that the disease was widespread, whereas the combatant nations were unwilling to admit that anything was wrong. More than 50 million people died as the pandemic spread across the globe. It killed around 5% of those infected, compared with the death rate of only 0.1% associated with seasonal flu. At the start of the outbreak, the flu was relatively mild but it soon mutated into a much deadlier form which was a strain of sub-type H1N1. Another, but less deadly, pandemic began in China in 1956 and as it spread around the world it killed as many as 4 million people. It was known as Asian flu and was sub-type H2N2. A potentially deadlier form of the virus, bird flu, officially known as H5N1, appeared in 1997 but thanks to worldwide cooperation this was contained mainly to East Asia and only 200 died. Another pandemic, swine flu, began in Mexico in 2009 but happily it is not very deadly.

It is no good taking an **antibiotic** to treat a viral disease. Antiviral agents are required, of which there are many although relatively few are prescribed. The best known one is oseltamivir, better known as Tamiflu. This is now stockpiled in tens of millions of doses. In a serious epidemic it would be delivered to homes in response to on-line or over-the-phone diagnosis. The dosage is 75 mg capsules taken twice a day for five days while the disease runs its course. Children under 13 should be given only one tablet a day because of side effects. In Japan, where Tamiflu is widely used, children have been reported to suffer behavioural problems and hallucinations, although the manufacturer, Roche, believes that these may be symptoms of the disease itself rather than the treatment. (The main side effect of Tamiflu is a feeling of nausea.) In the US the FDA has approved Tamiflu for children aged one and over. Treatment should begin within two days of the first symptoms appearing, or it can be taken to prevent the disease by those working in key locations like hospitals.

So how does Tamiflu work? The drug interferes with the final stage of a virus invasion of a cell, when the virus seeks to escape and for which it needs neuraminidase, an enzyme that sits on its surface and which it uses to bore its way out of the cell by partly dissolving the cell wall. Tamiflu itself does not block neuraminidase but it is converted to the active form when it is hydrolysed in the liver.

Tamiflu is manufactured by a sequence of chemical reactions starting with shikimic acid. This is extracted from the culinary spice star anise, which is grown in China and Vietnam, and at one time it looked as if this would be the limiting factor to producing Tamiflu, but production at

Roche is now around 400 million treatment courses a year. Making shikimic acid commercially is a ten-step process, although research has shown that this can be simplified.[4] An alternative method of production has been to genetically modify *E. coli* to make the acid.

Tamiflu-resistant strains of the flu virus can develop, which they do by mutating their neuraminidase enzyme, but these resistant viruses are much weaker at infecting other cells. In fact, the gene which generates the enzyme is not very versatile and since almost all viruses use the same enzyme, mutations are not likely to be Tamiflu resistant.

In addition to Tamiflu there are other antiflu drugs, such as amantadine, rimantadine and zanamivir. Like Tamiflu they will reduce the duration of the illness from around a week to around five days. These drugs, with the exception of zanamivir, are also recommended to be taken as a preventative measure by those exposed to a viral disease.

Another way to interfere with a virus that is trying to replicate itself is to block the DNA polymerase enzyme it needs to convert viral RNA to viral DNA, and there are drugs which do that, such as nevirapine (aka Viramune). Nevirapine was approved for use in the US in 1996 and in the EU the following year, but it can damage the skin to such an extent that it is life-threatening in 1% of those treated with it, and its use is now restricted.

It is also possible to defeat a virus by tricking its RNA into inserting the wrong kind of nucleoside when constructing its DNA and to do this by offering something which looks very similar to the molecular components it needs. One drug which does this is zidovudine (aka Retrovir and AZT) and this is used to control HIV. This drug had been made in the 1960s and was intended to be an anticancer drug but failed in its trials. Then when HIV, and the AIDS it causes, was recognised in the 1980s, zidovudine was looked at again and proved to be effective and was the first drug to be approved for HIV treatment in 1987.

There are many other antiviral drugs and many other viral diseases, some of which generally have a fatal outcome such as dengue fever, ebola, and West Nile virus. Maybe one day pharmaceutical chemists will find a drug to defeat even these deadly agents. Meanwhile the most common viral disease we encounter is the common cold and this poses such a low risk that treatment is mainly to deal with symptoms. (Other viral diseases like mumps and measles are best protected against by immunisation.)

[4] Barry Trost's research group at Stanford University has found a way of making Tamiflu in only 8 steps with an overall yield of 30%.

3.4 MULTIPLE SCLEROSIS (MS)

This disease of the nervous system blights many lives but help is coming in the form of old drugs with new tricks.

MS is a disabling disease which leads to loss of balance and muscle control, and then to paralysis and loss of memory, and eventually death. My sister had MS for more than 20 years and died in January 2008. About one person in a thousand has MS, and the cause appears to be a genetic predisposition and exposure to something which triggers the immune system into action. The immune system is primed to recognise invaders, and then its white blood cells will kill them. In the case of a virus, it will identify the enemy as an alien from protein structures on its surface. The immune system may then go on to misidentify and attack similar proteins on the surface of the myelin sheath, which is the fatty layer that acts like insulating coating for the nerves.

Existing drugs to treat MS, such as Betaseron and Avonex which are based on interferon, have to be injected. Interferon is a protein produced by cells that have been invaded by a virus in an attempt to stop the virus multiplying. These treatments tended to leave the patient with symptoms associated with flu for a few days. There are other drugs but these are often compromised by unpleasant side effects. In any case, these treatments tend eventually to be less and less effective. The reason for this may be that they cannot prevent deterioration of the nerve fibres and it is this which becomes the most serious part of MS.

Now there are several small molecule drugs being tested which can be taken orally. Some of these are based on drugs that were researched many years ago but were rejected because of side effects. They have been looked at again and ways have been found of modifying them so as to eliminate side effects. The pharmaceutical company Novartis is looking at fingolimod which is based on a fungus extract used in Chinese medicine for thousands of years. Fingolimod controls the release of white blood cells and is in the final stage of testing.

Another drug, cladribine (aka Leustatin), suppresses the white blood cells completely. It was designed to treat a rare form of leukaemia in which the white blood cells are misbehaving, and it tricks its way into the DNA of these white cells and makes them unable to replicate. (Normal cells spot it for the impostor it is and destroy it.) Cladribine is already approved for the treatment of leukaemia and for MS patients. For the latter, it is given in a short course of 5 days treatment twice a year. In leukaemia cases it is injected, whereas for MS patients it can be given orally. Cladribine is seen as potentially so successful that it is being fast-tracked by the FDA in the US and should be available in 2010. Those given the drug for two years saw the number of relapses greatly reduced and many were even relapse-free during this period.

Teriflunomide is Sanofi–Aventis's MS drug and is also in the final stages of testing. The drug moderates white blood cells rather than suppressing them. The drug was detected in the body of those given another drug, leflunomide, for rheumatoid arthritis, which is also an auto-immune disease, and it was working because it was being transformed to teriflunomide in the liver. Now patients can be given teriflunomide directly. This molecule targets an enzyme which is part of the immune system but doesn't completely knock it out.

Two drugs which appear to be even better and gentler are laquinimod (from Teva) and BG-12 (from Biogen). The former is based on an earlier drug which failed its final tests in 1997 when it appeared to increase the risk of heart attacks. Now it has been subtly modified in order to remove this risk and is again in the final stages of testing. The other drug, BG-12, is likewise an adaptation of a drug once used to treat psoriasis, and this newer drug works by actually protecting nerve fibres against attack. BG-12 is dimethyl fumarate and it has to be injected. Rather bizarrely, dimethyl fumarate is banned for use in consumer products in the EU after several people were affected by it when it was used to protect sofas from mould while in transit from China. They developed severe eczema merely by sitting on a sofa down the back of which was a small sachet of the chemical.

Fampridine is 4-aminopyridine which is used to poison troublesome birds and sold to farmers as Avitrol. It kills by blocking potassium channels in the nervous system. However, in people with MS this actually improves the transmission of nerve impulses along damaged nerve fibres, and those who have been given small doses of the drug report not only improved mobility but improved eyesight, and in tests more than 80% of those using it reported long-term benefits. It is prescribed in small non-toxic doses of 20 mg and given in the form of slow-release tablets.

MS can be difficult to diagnose in its early stages but magnetic imaging and blood tests will confirm it, and clearly the sooner these tests are carried out the better, although if the symptoms, such as tingling in hands and feet, are only temporary they may be ignored or attributed to some other cause.

3.5 INFECTIONS

Bugs like MRSA appear to be winning but chemists are fighting back.

Humans have an inbuilt defence against invading pathogens in the form of short-chain protein molecules (peptides) made up of between 15 and 20 **amino acids**. However, some bacteria have long since developed resistance to them, hence the need for **antibiotics**. These have to be strong enough to do the job and yet not damage healthy cells. Researchers are now trying to find other short-chain peptides against which the bacteria have no defence. The problem is to ensure that the peptides don't attack our own cells, such as red blood cells, which have also learned to live with the natural short-chain peptides.

In 1935 the first antibacterial drugs which could be taken internally were introduced and these were the sulfonamides which were able to cure infections, such as blood poisoning, childbed fever and erysipelas. Alexander Fleming had discovered penicillin, the first antibiotic, in the late 1920s but it was not until 1941 that it was shown to be effective in humans and very soon it was curing a wide range of deadly diseases, such as pneumonia, meningitis and syphilis. Fleming ended his Nobel Prize lecture in 1945 by predicting that bacteria would one day develop resistance to penicillin. That it would happen quickly was not expected, but by 1952 more than half of all *Staphylococcus aureus* infections were penicillin-resistant.[5]

The 1940s also saw cephalosporin introduced, which had the same key molecular feature as penicillin, and also streptomycin which was able to cure tuberculosis although treatment for many months is often necessary. Streptomycin kills microbes by inhibiting protein synthesis which eventually leads to the death of the tuberculosis bacteria, and it does not affect human cells because our cells make protein in a different way.

In 1950 a new class of broad-spectrum antibiotic drugs appeared, called the tetracyclines, and that decade saw yet more new antibiotics, including vancomycin which had been discovered in a soil sample from Borneo. Vancomycin was an extremely powerful antibiotic and became known as 'the antibiotic of last resort'. In the 1960s, the pace of discovery slackened because there seemed little point in finding yet more antibiotics when the ones that were available were performing so well.

In the past 40 years only two new types of antibiotic have appeared: the oxazolidinones and the **lipopeptides**. Approvals for new antibiotics in

[5] In 2005 it was discovered that penicillin might have a role in medicine yet again because it was found to be beneficial in treating degenerative diseases of the central nervous system, such as Huntingdon's Chorea. It does this by removing excess glutamate from the brain.

Figure 3.2 It's surprising where new drugs are to be found.

the US in the mid 1980s were on average three a year but this has dropped to less than one a year, and some quickly lose their power. Pfizer's linezolid (aka Zyvox), was the first entirely new type of antibiotic to appear in 35 years when it was marketed in 2000. It disrupts bacteria by blocking their ribosomes which they need in order to make proteins. Sadly some strains of bacteria developed resistance to linezolid within a few months of its use.

A promising new antibiotic, daptomycin, was discovered in soil collected on Mount Ararat in Turkey and was researched by the drug company Eli Lilly in the 1990s. The drug disrupts DNA, RNA, and protein production within the bacterium and kills the cell. It worked well, but when tested with large doses on volunteers they reported hurting muscles and so it was shelved. In any case Eli Lilly, like all the major pharmaceuticals companies, was getting out of antibiotics because research was slow and the results unpredictable. Nevertheless, they licensed daptomycin to a smaller company, Cubist Pharmaceuticals, which is based at Lexington, Massachusetts. Cubist Pharmaceuticals discovered that the side effects could be avoided if the drug was given as a large dose but less often. Cubicin, the new name for daptomycin, was launched in 2007 and in its first year generated around $300 million of income. The success of Cubicin has spurred other smaller biotech companies to develop new antibiotics and several of these could well be launched in the next few years.

Methicillin-resistant *Staphylococcus aureus* (MRSA) is the bug which everyone in hospital dreads. It appeared in the 1990s, and by 2007 there were 6300 serious infections due to it in the UK of which 1590 resulted in death.[6] It is particularly deadly to a person with an impaired immune system. Methicillin was a powerful antibiotic discovered by the UK pharmaceutical company Beecham in 1959. It was hailed as the drug which would kill bacteria that had become resistant to penicillin, and for a time it appeared that nothing could resist its power, but now MRSA can.

Common Sense 9: Antibiotic-resistant bacteria like MRSA are a threat to us all

Wrong. Although people whose immune system has been suppressed, for example after transplant surgery, are particularly at risk from antibiotic-resistant microbes, the rest of us are not. Alarming newspaper reports in the UK of finding MRSA on door handles in hospitals were proved false. The laboratory of the 'expert' to whom journalists had sent their samples turned out to be his garden shed.

MRSA and indeed other resistant bacteria do pose a serious threat to people in hospital, but if the rest of us take simple hygiene precautions then we are not in danger. In fact, the general public are more likely to introduce dangerous pathogens into hospitals when we visit sick patients or newborn babies. The best preventative is still the old advice: always wash your hands before you touch food and after you have been to the toilet.

Bacteria can survive deep within the Earth's surface, in acidic lakes, at the temperatures of near-boiling water, and under the most extreme cold. Given their remarkable versatility it is not surprising that they dominate life on this planet and were the only forms of life for a billion years or so. As far as we are concerned, most of them are benign but a few are not, and it is those which can eventually become resistant to any antibiotic that chemists discover. It isn't a case of *whether* a pathogenic bacterium will become resistant, rather a case of *when*.

[6] In the US about 2 million people get hospital infections every year and 90,000 die as a result. The WHO estimates that 60% of hospital-acquired infections are now drug-resistant.

The reason that antibiotics are sought in soil samples is because this environment is full of bacteria and yet other forms of life have developed ways of coping with them, and indeed some bacteria themselves have developed chemicals that kill other bacteria. For example, the pharmaceutical company Merck has found platensimycin in soil from South Africa, and it is produced by the bacterium, *Streptomyces platensis*. It interferes with the enzyme that most bacteria use to make an essential fatty acid. Platensimycin is effective against *Streptomyces aureus* and even against MRSA.

There are various ways to attack a bacterium: at the point when it is trying to make its cell membrane; when it is trying to make new proteins; when it is synthesising DNA and RNA; and when it is making precursor molecules for making DNA and RNA. Bacteria fight back. They protect the site which the antibiotic targets; or they find a way to identify the antibiotic and reject it before it can damage the cell; or they may develop a better cell membrane which the drug cannot penetrate.

Another weak point in the life of a bacterium is when it divides and needs a special protein to enable it to do so. Chemists at a small company, Prolysis, based at Yarnton in Oxfordshire, took the compound 3-methoxybenzamine, which will interfere with the bacterium at this stage ***in vitro*** but not ***in vivo***, and they then made 500 variants of it until they came up with one which worked *in vivo*. Whether it will make it to the market place remains to be seen.

Bacteria like *Staphylococcus aureus* cause nasty food poisoning, as well as being a source of hospital infections. So far they have failed to become resistant to vancomycin but that is almost certain to happen one day.[7] Vancomycin is an antibiotic which works by interfering in a bacterium's ability to make cell walls. What Brendan Crowley and Dale Boger of the Scripps Research Institute in California have identified is a way in which vancomycin could be modified to make it an even deadlier weapon against the construction of bacterial cell walls. In 2002 Boger was also instrumental in finding a way of making the powerful antibiotic ramoplanin. This molecule was extracted from a strain of bacterium *Actinoplanes*. Ramoplanin is now available for treating infections of the gut and especially *Clostridium difficile*. Again it works by disrupting the enzymes needed by bacteria for making cell walls.

[7] It has already happened but has been contained. In July 2002, doctors in Michigan were treating a man with an ulcer on his foot and discovered the first example of vancomycin-resistant *S. aureus*. The bacterium had been living in the wound along with a harmless bacterium *Enterococcus faecalis* which had acquired vancomycin-resistance and with which it had exchanged genes.

In the past the tendency has been to use so-called 'broad spectrum' antibiotics which will kill all bacteria, good and bad, but today some companies are now concentrating on narrow-range antibiotics that target the bacterium known to be causing the infection. Such a drug has been discovered which kills the *Clostridium difficile* that colonises the lower intestine and leads to a particularly persistent and damaging kind of diarrhoea. It has developed resistance to many drugs and it kills more than 1000 patients a year in UK hospitals. Hospital labs reported more than 43 000 cases of this infection in 2004. The new drug is being researched by the US company Optimer and will be available from 2010 onwards – if tests are successful.

3.6 CHEMOTHERAPY

Cancer was once a death sentence. Now it can be fought and even cured.

Chemotherapy is saving lives, and even those with incurable cancer are living much longer. The first line of attack is generally surgery and radiation treatment, followed by powerful drugs to target the remaining cancer cells which by their very nature are rapidly dividing. The drugs also affect other rapidly dividing cells, such as hair follicles, bone marrow, and the lining of the gut, resulting respectively in loss of hair, fewer red blood cells, and vomiting and diarrhoea. Clearly a course of chemotherapy is not going to be pleasant but at least there is now a good chance it will add years to a patient's life and might even rid the body completely of cancerous cells.

So what molecules have chemists devised which can attack cancers and how do they work? One particularly effective group are the alkylating agents and the first of these was mechlorethamine, which is the medical name for what was once known as mustard gas, the chemical warfare agent. Another anticancer drug is the platinum compound cisplatin, which is a simple inorganic chemical consisting of a central platinum atom to which are attached two chlorine atoms and two ammonia molecules. It is called *cis* because the whole molecule is planar and the two chlorine atoms are next to each other, as are the two ammonias. If they were on opposite sides then the molecule would be the *trans* arrangement and while cisplatin can cure testicular cancer, transplatin cannot. Another platinum-based drug is oxaliplatin (aka Eloxatin) which has different molecules attached, and these must adopt a *cis*-type by their very nature, and this too has proved useful in the treatment of colon and rectal cancer when used in conjunction with other drugs. More than half of those treated with it are still alive after five years, and not surprisingly it is a blockbuster drug.[8]

The cisplatin story began in the late 1960s at Michigan State University where research was going on into the effects of electricity on living cells. It was observed that filaments of dead bacteria formed on the platinum electrodes which were used. A little platinum was dissolving and reacting with chlorine and ammonia in the growth medium to form cisplatin, and it was this which was stopping the bacteria from dividing and making them cluster into filaments. Could it do the same with cancer cells in the body? The answer was yes, but it would do the same to all cells, so treatment had to be localised to avoid damaging the kidneys. As far as testicular cancer was concerned it was superb with a 100% cure rate. That success spurred others to make similar platinum anticancer drugs and today there are two more potent ones: carboplatin and oxaliplatin. These are used to treat ovarian, breast, lung and neck cancers. Other metal-based drugs are being researched and not only based on platinum but on metals, such as titanium, gallium, ruthenium and osmium, and some of these are already able to shrink breast cancers by up to a third.

What drugs of this type do is interfere with the cancer cell's DNA by reacting chemically with its components so it cannot replicate properly. Some chemotherapy drugs can trick a cancerous cell with a chemical that looks like the one it needs to make DNA but which in fact has been subtly modified. The cell incorporates the false molecule into the DNA strand it is making, only to find later on that it has taken in a poison pill.

Another approach is to interfere with the cancer cell as it starts to divide and for this the cell needs microtubules, which are tiny rods of diameter around 25 nm and up to 25 μm long. It needs these to maintain its shape, so that preventing these from functioning makes it impossible to form a new cell. There are various plant-derived chemicals which have been found to do this, such as the vinca alkaloids from the Madagascar periwinkle, podophyllotoxin from the rare Himalayan mayapple, and paclitaxel (brand name Taxol) from yew trees. Having found these active agents, chemists devised ways of making them in the laboratory and changed them slightly either to make them more effective or to reduce side effects.

How an anticancer drug is given depends on its toxicity, with some having to be injected directly into the cancer, as with liver and lung cancers, because they are too toxic to be delivered *via* the blood stream. This type of treatment might involve a highly radioactive isotope which will kill the cells around it when it emits its damaging radiation.

[8] These are drugs whose sales exceed $1 billion a year and there are about a hundred of them.

Nanoparticles might also be used to deliver a dose of chemotherapy either because they can transport drugs which are not very soluble, or they can carry the drug and a tiny molecular magnet which can be used to direct the nanoparticle to the cancer by means of an external magnetic field.

Sometimes a natural molecule is found which has powerful anticancer properties and one such was leucascandrolide A, which was first isolated in 1996 from a sponge collected from the Coral Sea.[9] However, when more of this sponge was collected it yielded no more of the chemical, and it appeared that it probably came from a bacterium living on the sponge. The challenge then was for chemists to make this complex molecule in the lab and that was achieved in 2008 by Andrew Evans at the University of Liverpool and he constructed it in 14 separate stages.

Diagnosing cancer early on would be a great benefit to patients, and one way this might be done is to detect molecules that are being produced by it. It is known that dogs can recognise certain kinds of skin cancer by smell and what chemists at the Monell Chemical Senses Center in Philadelphia, Pennsylvania, have done is to analyse volatile compounds being emitted by the skin of cancer patients. What has so far been found is that skin cancer cells release more dimethylsulfone and less 6-methyl-5-hepten-2-one than healthy cells and that this change can be detected even before there is physical evidence of a tumour growth on the skin's surface.

Some chemotherapy can be remarkably successful in stopping a cancer in its tracks – but only for a while. Imatinic, marketed as Gleevec, is one such, but it eventually stops working because the cancer learns how to cope with it or it finds an alternative way to spread. Just as we observed with antibiotic-resistant bacteria, so can cancerous cells develop their own defence against drugs. They may reject the drug by pumping its molecules out through tiny channels in the cell wall, so-called efflux pumps. Drugs which can block these pumps are therefore needed and some are currently on trial.

What chemists are now looking for are drugs which target a receptor known as *Met*, which adds a phosphate group to a tyrosine molecule that is part of a protein and which is implicated in many cancers. *Met* has various roles in the body, including in the functioning of bone marrow; in the repair of a liver with cirrhosis; and as a key player in wound healing. However, it also helps cancer cells remain alive even

[9] Called *Leucascandra caveolata*.

after they have been damaged by a drug. Pharmaceutical companies have now found drugs that will control *Met* and these are currently being tested.

What would be particularly helpful for both clinician and patient would be to know how a person will respond to a particular chemotherapy drug. Normally there is a 50 : 50 chance that a particular drug will work but for some drugs the success rate is only 20%. A University of Manchester spin-off company Oncoprobe hopes to increase the success rate by testing biopsied samples of the tumor. These will attach themselves to a gold probe and their response to a drug can be monitored with a tiny electric current. Whether this discovery will one day be part of the standard diagnosis of cancer remains to be seen.

3.7 ASTHMA

Gasping for air can be relieved very easily.

As we breathe, air passes down the windpipe (aka the trachea) to the bronchi and thence to the alveoli, where in tiny air sacs its oxygen molecules can diffuse into the blood stream while carbon dioxide can diffuse out to be expelled from the lungs as when we breathe out.[10] Sometimes, and for no obvious reason, the tubes of bronchi suddenly narrow and then that person is gasping for breath and an asthma attack is underway. In earlier years if the attack became severe then medical treatment would be needed and an injection of **adrenaline** given. Today when an asthma attack begins, the person reaches for their inhaler and one puff of salbutamol (aka Ventolin) brings relief that lasts for several hours. What this drug is doing is delivering a modified kind of adrenaline to the site where it is most needed.

Adrenaline acts as a neurotransmitter, the action of which is to shift bodily processes up a gear. (This is the fight or flight hormone.) Most noticeably the heart begins to pound and the adrenaline helps to pump a greater volume of blood around the body, especially by dilating the blood vessels to the muscles and liver, the latter which helps release glucose from its energy store. Adrenaline also dilates the pupils of the eyes. Once the threat is over, the excess adrenaline disappears and our heartbeat returns to normal.[11]

[10] The lung muscles responsible for drawing air into the lungs are very powerful, but those whose role is to expel the air are much weaker.

[11] The body deactivates adrenaline *via* the action of *monoamine oxidase* enzymes, and then discharges the resulting chemical into the urine.

Figure 3.3 Breathe easily again with a tiny puff of chemistry.

In the lungs, adrenaline relaxes the walls of the bronchi and breathing becomes instantly easier. Originally the adrenaline had to be injected so medical help was required. Adrenaline injections gave way to injected steroids which were much more effective, but what was clearly needed was a way of giving relief without injection, which the sufferer could use, and which lasted for several hours. The answer was found at Allen & Hanbury[12] in the UK, where research chemist David Jack discovered salbutamol. This did all that was required and its effects lasted four hours. It first appeared in 1969 and then was launched in the US in 1980 under the name by which it is now widely known: Ventolin. Moreover, salbutamol was easily made from aspirin.

Of course salbutamol is not a *cure* for asthma – nothing has been discovered which can do that. Persistent sufferers can be treated with a steroid whose inhaler comes in a brown casing as opposed to the blue Ventolin one. The steroid acts as an antiinflammatory agent but it takes several days to have an effect although when it does it lasts a long time. There are now other drugs which act in the same way as Ventolin and these are salmeterol, which is prescribed for severe and persistent asthma, formoterol, which is also prescribed for chronic pulmonary

[12] This became part of what is now GlaxoSmithKline in 1958.

disease, and bambuterol, which is a longer-acting asthma drug. Combinations of inhaled steroids with these kinds of quick-relief drugs are now available.

Another condition in which the lungs fight for air is cystic fibrosis and there are 8000 people in the UK and 30 000 people in the US with this disability. It is caused by an errant gene inherited from one's father and mother and it is a life-threatening condition in which the lungs become clogged with thick mucous. This has to be removed mechanically by repeatedly compressing the chest. That can be done with a special vest, while the person inhales a saline spray which thins the mucous and helps it to clear. What threatens the health of those with this condition are bacteria living in the lungs, and in particular the deadly *Pseudomonas aeruginosa* which has to be kept in check with hefty doses of antibiotics.

Back in the 1950s, those born with cystic fibrosis rarely reached school age. Then with the development of powerful antibiotics by the 1980s they could expect to live almost to their late teens, and with modern medication their average lifespan has doubled so that most reach their late 30s. A breakthrough in understanding the condition was a discovery in 1989. A particular gene, called cystic fibrosis transmembrane conductance regulator, was found to have mutated which impaired its role in making the protein which moves chloride ions across membranes. Those with cystic fibrosis have too few 'gates' through which chloride can pass and this has most effect in the lungs where it means very little fluid oozes into the airways.

Knowing what the target gene was like, enabled chemists to search for drugs that might create more 'gates' on the surface of lung tissue, or boosting the action of the few which do exist so they transport more chloride and water. Several promising molecules have been found. Some modify the defective gene while other drugs target the movement of chloride ions. These drugs are now at advanced stages of testing. There is every hope that within a few years cystic fibrosis will be disease that people can live with rather than die from.

Seven areas in the field of healthcare are presented here in which the work of chemists directly affects the lives of millions of people. Of course, there are many other areas of life where pharmaceutical drugs promote healing and longevity. And yet the industries which produce these drugs are often the target of ill-directed comment by those claiming that their profits are too large; their profits are generated at the expense of the sick; and that their drugs are priced so that few in the developing world can afford them. The problem is that drug research requires a great deal of money and a large number of highly trained chemists.

Of 100 000 potentially useful molecules for treating a medical condition, only one will make it to market after years of testing and waiting for official approval. Even then the results may be costly for the company. Merck had a new treatment for arthritis, Vioxx, which unlike other such drugs did not cause stomach ulcers, although it was found to be associated with a slightly higher death rate among millions taking it.[13] Although relatively few were affected, it was enough to trigger a series of scare stories in the media and a massive class action in the US. Eventually the company had to pay more than $4 billion in compensation, and Vioxx is no more.

Of course those in the industry are not entirely blameless, and we should not be too surprised when we find them not behaving like charitable organisations. They have shareholders, like pension funds, who demand dividends. The pharmaceutical industry is huge with sales amounting to around $289 billion per year (2007).

The top five companies are:

Pfizer (US, sales $26b)
GlaxoSmithKline (UK, sales $20b)
Merck & Co (US, sales $17b)
Johnson & Johnson (US, sales $16b)
AstraZeneca (UK, sales $15b).

In fact all are international in their operations. Much of their income comes from blockbuster drugs. The five best-selling ones currently are:

Lipitor (Pfizer, $8b), anti-cholesterol statin
Nexium (AstraZeneca, $5b), heartburn drug
Advair Diskus (GlaxoSmithKline, $4b), asthma treatment
Plavix (Bristol-Myers Squibb and Sanofi-Aventis, $4b), heart attack preventer
Seroquel (AstraZeneca, $4b) anti-depression treatment.

Is there a sustainable future for the industry? The answer is yes, because the quantities of chemicals needed are relatively small so that when only renewable resources are available they will still be produced.

[13] In fact, this increased mortality had been known all along and accepted as a risk by the FDA and NICE in the UK. It was probably due to the fact that Vioxx users no longer had to take arthritic pain-killers, such as aspirin which has a known protective effect against coronary artery disease.

Some can indeed be extracted from genetically modified plants although even this might not be necessary. For example, the antibacterial agent carvacrol is found in oil of thyme but the oil from the Himalayan oregano contains as much as 75%. A chemically modified version of this molecule, to make it more powerful or better targeted, might one day be part of the medical pharmacopoeia.

Later this century things will change and new drugs treating other diseases will top the lists and there are already indications of how these might come about. For example, pharmaceutical chemists have discovered a new way to make drugs more effective and that is to replace some of the hydrogen atoms in their molecules with heavy hydrogen (deuterium). In some cases this makes them last longer in the body, in other cases it actually makes them work better. Two such modified drugs are currently being tested, one of which is used in the treatment of HIV, the other in the treatment of kidney disease.

CHAPTER 4

Transport Biofuels and Chemistry

[A word in **bold** means there is more information in the Glossary.]

Seven biofuels will be transporting us this century – possibly. Two are already being used: bioethanol and biodiesel. The others are biobutanol, biomethanol, biogasoline, and the gases biomethane and biohydrogen. No biofuel is yet without its drawbacks, some because the crops that produce them compete with food crops, others because they are gases, such as biohydrogen.

The thought that **fossil fuels** might one day run out has worried humans for more than 125 years. In 1875 the father of science fiction, Jules Verne, wrote in his novel, *Mysterious Island*:

'... and what will men burn when there is no coal? Water. Yes, my friends, I believe that one day water will be employed as a fuel, that hydrogen and oxygen which constitute it, used singly or together, will furnish an inexhaustible source of heat and light.'

This century, his prediction might well come true. Of course we won't be using water as such, but if our children and grandchildren really can harness science and technology to solve the world's energy needs in a sustainable way, then we might well be extracting and burning the hydrogen it contains. Maybe it will even be used to transport us – at least to the Moon, if not to our local supermarket.

A Healthy, Wealthy, Sustainable World
By John Emsley

Published by the Royal Society of Chemistry, www.rsc.org

Common Sense 10: We could run cars on water but the oil companies keep it a secret

Wrong – and right. Of itself, water has no energy left to give, so the idea of putting water in the fuel tank of a car, along with some magic ingredient which oil companies know about, is merely an urban myth. However, as a source of hydrogen gas, water has the potential to be a fuel of the future. Hydrogen is produced when steam reacts with biomass, or when electricity is passed through water, or even when sunlight falls on water containing the right catalyst, but in this last case only in tiny amounts. Hydrogen gas generated from water can be used as a fuel for transport, and vehicles have been produced which run on it.

When the world's fossil fuel reserves are no longer available then it will be up to Nature to provide the raw materials to fuel our cars, trucks, tractors, ships and aircraft. Can we really have a world without fossil fuels and one where everyone enjoys the mobility we associate with a developed lifestyle? For this to happen we need to produce renewable fuels. According to BP's *Annual Review of World Energy 2009*, we consume almost 4 billion tonnes of oil a year, which is 2% of the known reserves of 195 billion tonnes. Half of the oil goes into transportation fuel. Clearly replacing all that with fuel derived from harvested crops will never be easy, yet the annual biomass production on land is more than 50 times the weight of oil we use as fuel, which shows that in theory it is possible. It's a simple as that – and it's as difficult as that.

Sustainable transport fuels need to be high-energy liquids, and these must derive from **biomass**, a term which encompasses all kinds of plant matter from ten-metre tall trees to incredibly small algae. We know biomass is rich in energy because some forms of it, such as logs, can simply be dried and burnt to release it. We could even return to the days of the steam engine and burn logs in the furnaces of boilers to drive trains and ships. That would still leave the problem of how to fuel road vehicles and aircraft. For the engines of these we need to convert biomass to liquid fuels – and that's where chemists enter the picture.

The scale of the problem is daunting when we consider air transport, and this was demonstrated by a Virgin Atlantic Airway's Boeing 747 flying from London to Amsterdam in 2008. It consumed 22 tonnes of fuel, of which 1 tonne was biofuel.[1] Those who hailed this as the first

[1] It had to be diluted with ordinary fuel to stop it congealing at the low temperatures of high altitudes which biofuels are prone to do. The Finnish oil company Neste has produced a biodiesel called NExBTL which can be used by aircraft flying at great heights.

step into a future of sustainable air transport were brought down to earth when it was pointed out that the biofuel component of that flight consumed as much oil as 150 000 coconuts could produce. Had all the fuel been biofuel then it would have required the oil from 3.3 million coconuts – and for a journey of only 356 km. Aviation consumes 340 billion litres of jet fuel per year. If all this were to come from plant oils, such as palm oil, then it would require around 1.4 million square kilometres to be planted, which is twice the area of Texas or nearly 15 times the area of the UK. The Royal Society of Chemistry calculated that it would take 30 football pitches (about 20 hectares) to produce the 80 tonnes of kerosene required for a single flight from London to New York. Clearly fuelling all of the world's aircraft with biofuels is not going to be easy.

Plant biomass is roughly 50% **cellulose**, 30% **hemicellulose** and 20% **lignin**. These various components are resistant to being broken down, so their energy is hard to extract. Plants developed this material to be tough and resistant to decay, but it will one day have to provide the fuels we need. On the other hand, the food components of plants offer easier sources of energy, such as sugar, starch and seed oils. The most energy-rich plant components are the **triglyceride** oils, and while many plant oils are seen as food crops, there have always been some which were better burnt in lamps to provide lighting. Animals also produce triglycerides in the form of fats, and while these too were regarded as food, some was also used for lighting, in this case as candles.

The basic **unit** which science uses when talking about energy is the joule, an amount which is little more than that required to walk a few steps. Chemists prefer to use the kilojoule (kJ), which is a thousand joules, but even this amount would only increase the temperature of a small beaker of water by a mere one degree. A human being needs around 8 000 kJ per day. Given such an input of energy we can easily walk 10 miles (16 km). To drive this distance in a car we normally need to burn a litre of fuel and this would consume 70 000 kJ of energy. Clearly cars are very inefficient in terms of transporting us around. However, if a car could travel 50 miles (80 km) on a litre of fuel it would be a big step towards sustainability, and if it could travel 100 miles (160 km) then the private motor car may have a permanent future. That thought might elate you – or depress you.

The comparison of one form of fuel with another is difficult when some are liquids and some are gases. For example, a gram of hydrogen gas releases more energy (143 kJ) than any fuel, but it occupies an unwieldy 11 litres. By comparison, a gram of natural gas (methane) releases 56 kJ per gram and occupies 1.4 litres. Liquids occupy much less space. A gram of petrol occupies only 0.012 litre (1.2 ml) and releases 48 kJ, about the same as a gram of diesel. Consequently cars of the future are still likely to run on liquid fuels, and this means hydrocarbons, alcohols or esters. Currently we use **hydrocarbons** derived from oil and these consist of short chains of carbon atoms.

Were cars to become more efficient users of energy then the average car might need only 200 litres a year, or less. Some of this might even derive from municipal waste. A city of 1 million residents will throw away about 100 000 tonnes of organic matter a year and this could in theory be converted to 15 million litres of fuel, enough to supply 75 000 cars with 200 litres per year. Such a city would have many more cars than this, but waste-into-biofuel might well power around 20% of them.

Global biofuel output in 2006 was 40 billion litres, of which bioethanol accounted for 90% and biodiesel for most of the remainder. These are called first-generation biofuels and are produced from crops normally grown for food. Second-generation biofuels will be produced from non-food biomass. This can be converted to high-energy chemicals like ethanol and butanol. Alternatively, biomass can be turned into gas by reaction with steam at high temperatures and this synthesised gas – or syngas as it is generally called – is a mixture of hydrogen and carbon monoxide, which can then be turned into biomethanol or biogasoline.

The overriding property of any transport fuel is the amount of energy it releases when burnt, but other properties are also important, such as its octane number which is related to the efficiency of burning (the higher this is, the cleaner it burns) and its melting point (too high and it might be solid on a very cold day). Indeed this happened in Norfolk in the winter of 2008 when a fleet of biodiesel buses were unable to run because their tanks were full of solid fuel. Viscosity is also important and the lower this is, the more readily a liquid will flow along pipe.

The five liquid fuels most likely to be the transporting us in the future are summarised in the following table:

Biofuel	*Energy (MJ litre^{-1})*	*Octane number*	*Melting point (°C)*	*Viscosity (cSt)*[a]
Methanol	16	135	– 97	0.64
Ethanol	20	125	– 114	1.52
Butanol	29	95	– 90	3.64
Petrol	32	95	– 57	0.5
Biodiesel[b]	33	n/a	– 10	ca. 5

[a]The viscosity of water on this scale is 1.
[b]Rape methyl ester.

Let us now look at individual biofuels and consider their advantages and disadvantages. The first two are already in use, namely bioethanol and biodiesel. The others which might join them are biobutanol, biogasoline, biomethanol, biomethane and biohydrogen gas.[2]

4.1 BIOETHANOL

This is a proven biofuel for cars and is becoming more widely used.

Henry Ford launched his famous Model T motor car in 1908 and he envisaged that it would be fuelled by ethanol. He told struggling farmers as late as the 1920s that this was the fuel of the future. He was wrong and he was right. At the time the fast-growing oil companies had a much better and cheaper product for the motorist and that was gasoline (petrol). Nevertheless, ethanol was one of the first gasoline additives, a few per cent of this chemical gave a cleaner burning fuel thereby preventing 'knocking', which was an indication that the fuel in the engine was exploding rather than burning in a more controlled manner. When air pollution from city traffic became a serious issue in the US in the 1970s, petrol stations began to offer customers 'gasohol' which is petrol with 10% ethanol and by 1986 ethanol production had reached 1.5 billion litres per year. Bioethanol, aka fuel ethanol, is made from sugar in Brazil, maize in the US and wheat in the EU.

Today there are more than 6 million cars on the roads in America which are able to run on E85, the fuel that is 85% ethanol and 15% petrol.[3] Cars burning E85 need filling more often but its price is a few cents lower than regular gasoline, although that depends on the region where it is available. Bioethanol is made from maize and US farmers can produce about 10 tonnes of this per hectare which yields around 1600 litres of bioethanol. Production has also been helped by allowing bioethanol-producing plants to be governed by the same laws as distilleries, and not to be classed as chemical plants with all the restrictions and extra costs which that implies. By 2008 there were about 100 biofuel refineries in the USA with a total capacity of 35 billion litres. The US Energy Independence & Security Act of 2007 has set an ambitious target of 220 billion litres of bioethanol by 2030. To produce this from maize would take double the amount of that crop already being grown in the US, leaving little for the other users. Currently only 20% of the maize crop goes into bioethanol production.

[2] There is now a dedicated journal dealing with biofuels: *Biofuels, Bioproducts and Biorefining* which is published by John Wiley on behalf of the Society of Chemical Industry, London.

[3] So far, however, only a few percent of the US's 170 000 petrol stations sell E85 and most of these are in the Midwest.

More of the current maize crop could be used to make second-generation bioethanol, by using the cellulose of crop residues, as we shall see.

The country which came closest to relying on bioethanol for its transport fuel was Brazil in the 1970s and 1980s. Brazilian farmers were able to increase yields of sugarcane to 77 tonnes per hectare, and processors were able to boost the amount of alcohol fermented and distilled to produce around 4800 litres per hectare. The cane is crushed, the juice filtered and then fermented with yeasts for around 8 hours to yield as much as 10% ethanol. The yeast is filtered off and some is sold as animal feed, the rest retained for the next batch of cane juice. The resultant liquor is distilled to extract the ethanol. The energy balance is very positive at 8 for 1, which means that for every unit of energy used in production, 8 units are created. The fibrous waste from the crushers, known as bagasse, is burnt in furnaces to produce steam and generate electricity, the surplus of which is sold to the local electricity company. Sugar cane is grown for a few years and then rotated with leguminous plants such as peanuts which replenish the nitrogen of the soil. Such agriculture appears to be sustainable.

In Brazil there are 'flex-fuel' vehicles which can run on petrol, ethanol or a mixture of the two. They are equipped with sensors which adjust the timing of the spark plugs so that the engine runs smoothly whatever mixture is used. Three quarters of their new cars are flex-fuel. In 2008, Brazil produced enough ethanol to power 45% of its passenger vehicles, from sugarcane grown on 3.4 million hectares which is a mere 1% of its arable land. This land is located in the centre of that country and not as popularly supposed taken from the Amazon rain forests, which are 2500 km further north. The country aims to triple ethanol production by 2020 and by only doubling the area of land needed to do so. A new processing plant at Edéia, Goias State, will produce 435 million litres of bioethanol a year.

In theory, a tonne of biomass should produce between 360–450 litres of second-generation bioethanol from cellulose as the raw material. Cellulose consists of **glucose** units, and has the potential to yield almost as much bioethanol as sugar or starch. Around 350 litres of ethanol could be made from a dry tonne of corn stover, which is the name given to the leaves, stalks and cobs of maize. This would be in addition to the 400 litres which can be got from the maize itself. The biofuel company Poet plans to have a plant in operation in 2011 at Emmetsburg, Iowa. Transporting biomass is not very efficient if it consists of mainly leaves and stalks, but cobs are much more economical to transport.

Past attempts to convert cellulose to bioethanol have been only partly successful. From 1976 to 1979 Gulf Oil Chemicals operated such a plant but it was not economical and the plant closed down. Today another company, Iogen based at Ontario, Canada, has a plant capable of producing 4 million litres of bioethanol a year from straw, which it does by treating crop residues with steam, enzymes and yeast. Similar plants are being built in Spain and Germany. This is the syngas route to ethanol, as opposed to the biotech route, and there is more about this in Section 3.4 below.

Converting cellulose to bioethanol by the biotech route can be done in three ways. One is to genetically modify plants so they have an inbuilt enzyme that would break down the plant's cellulose but would only be activated when heated. A second approach would also be to genetically modify plants, but to produce a modified form of cellulose that would be more easily converted to bioethanol. The third approach would be to improve an enzyme that is known to digest cellulose and make it more effective. Such an enzyme was extracted from the fungus found in elephant dung and it is a powerful digester of even the toughest plant material. It has been inserted into a yeast which will be used at a plant being built by Nedalco in The Netherlands. This will process 160 000 tonnes of food residues into bioethanol. The patented yeast can convert xylose, a major component of hemicellulose into ethanol.

Cellulose could be specially grown for conversion to bioethanol in the form of miscanthus grass,[4] switchgrass or poplar trees. Miscanthus (*Miscanthus giganteus*) grows quickly to reach 3 metres, and can yield 25 tonnes of dry biomass per hectare in a poor year and 40 tonnes in a good year. Switchgrass (*Panicum virgatum*) which grows in southeast USA can yield between 10 and 15 tonnes per hectare per year. Hybrid poplars are also a possible source of biomass because these are among the fastest growing temperate trees. They can reach a diameter of 20 cm and a height of 24 metres, and will yield up to 40 tonnes per hectare per year.

Cellulose which has found other uses, such as paper, could also be recycled to bioethanol. Research in the Department of Microbiology at the University of Florida, Gainesville, in the mid-1990s showed that it was possible. Lonnie Ingram and colleagues found the genes capable of doing this in a bacterium called *Zymomonas mobilis*, which is used by Mexican brewers to ferment the agave plant to make pulque, a weak alcoholic drink from which tequila is distilled. Because this bacterium did not survive in a mush of waste paper fibres, Ingram transferred the necessary genes into a hardier bacterium *Klebsiella oxytoca*, which flourishes in the waste water from paper mills – and it worked. BC International, based at Hingham, near Boston, Massachusetts, bought the licence to use the technology and they now have an experimental plant producing bioethanol from waste paper.

---

[4] Also called E-grass or elephant grass.

The fungus *Chrysosporium lucknowense* also contains an enzyme which can make bioethanol from biomass and this is being commercially exploited by the Spanish company Abengoa. The UK company TMO Renewables, based in Guildford, has discovered a thermophilic bacterium TM242 in compost. This has cellulose-digesting enzymes which can convert cellulose to bioethanol, and they have built a plant to operate this process.

Common Sense 11: If we use crops for fuel then there will be food shortages

Unlikely, but it is an issue that must be addressed. The world already produces more than enough to feed everyone and yet the poor go hungry while the affluent become obese. There is also the irony of poorer countries growing crops to produce biofuels for vehicles in developed countries when they could be growing food instead. Chemistry cannot solve political and social problems like these. In fact there need be little conflict between food and fuel once second-generation methods of producing the latter are available. Land unfit for farming could produce biomass for fuel.

When oil prices soared in the summer of 2008, much was written about biofuels but this prompted several counter articles pointing out their hitherto unnoticed drawbacks. Bioethanol and biodiesel were accused of using crops that could otherwise be sent as food to undeveloped countries, and of wasting water in order to grow these crops. Nor were biofuels as environmentally friendly as previously supposed, when the cost of transporting biomass to the processing plants was taken into account. To a certain extent these arguments are valid.

Nonetheless, there seems little point in not converting food to fuel if more is grown than can be consumed, and this applies even to an overcrowded island like the UK. Plants are now being built to turn surplus grain into bioethanol and surplus sugar beet into biobutanol. BP, along with other companies, is investing $400 million in a plant at Hull which will produce 420 million litres of bioethanol from one million tonnes of surplus wheat, and eventually 14 million litres of biobutanol per year from a smaller surplus of sugar beet.

Of course, if ever there were to be a global famine then producing fuel from food would be unthinkable.

The debate about the benefits of biofuels started more than 30 years ago. In 1973 David Pimentel, a professor of agricultural science at Cornell University, New York state, wrote a paper critical of bioethanol as a transport fuel and he has continued to criticise it, mainly on the grounds that when it is produced from maize it can never justify its cost. According to Pimentel, the energy it delivers is more than cancelled by the input of fossil fuels to produce it, to the extent of around 30% overall net energy loss.[5]

In 1987 a paper was published in *The Journal of Chemical Education* which showed how to calculate the energy needed to produce bioethanol from maize, and how to measure the energy that could then be derived from it. The conclusion did not bode well for this fuel. Only under the most favourable conditions was there a worthwhile 'profit' to be made. The energy to produce 1000 litres of bioethanol would be 20 000 kJ while the energy derived from it would be 44 000 kJ, yielding a profit of 120%. However, under normal conditions this profit was much less. Calculations have been refined since then and the US Department of Agriculture now says the likely profit is 67%. However, the Argonne National Laboratory says it is 35%, while Cornell University's researchers calculate only 27%, which is in line with Pimentel's findings. But even if the profit were as low as 20% it might still be worth doing for political reasons, at least in the USA where the aim is to reduce their reliance on imported oil.

4.2 BIODIESEL[6]

Biodiesel production now exceeds 95 billion litres a year, and rising.

In 2008 the New Zealand trimaran *Earthrace* broke the world record for a power boat to circumnavigate the world, travelling 24 000 miles in 61 days, which was 14 days quicker than the previous record. Its engine was powered by biodiesel.[7]

Before it left on its epic journey, its skipper Peter Bethune, a former oil engineer, contributed a tiny fraction of biodiesel made from his own body fat. He had 80 g of this removed by liposuction and it was converted to enough biodiesel to power his craft for 50 metres.

Diesel engines are more efficient than petrol engines and they fuel trucks, trains, tankers and tractors. The diesel engine was patented in 1893 by the German engineer, Rudolf Diesel.[8] It works on the principle

[5] Those who question his analysis point out that fossil reserves also require a large input of energy in order to produce fuel.

[6] Biodiesel is also known as FAME, short for fatty acid methyl esters. The fatty acids are palmitic acid, stearic acid, oleic acid and linoleic acid.

[7] The boat was built in New Zealand largely from renewable resources.

[8] Rudolf Diesel died under rather mysterious circumstances when he was on the steamer *Dresden* crossing the English Channel on 29 September 1913. He disappeared that night from his cabin and his body was found floating in the sea 10 days later.

Figure 4.1 Coming to a petrol station near you, but what's it made from?

that when air is compressed to less than 5% of its normal volume, its temperature rises to around 800 °C and is high enough to ignite any fuel that is injected into it. (Petrol engines rely on a spark plug to ignite a mixture of air and fuel.) Rudolf Diesel demonstrated his new engine at the World Fair in Paris in 1900 and it ran on peanut oil. At the time it seemed likely that plant oils would be the fuel for such engines, but again this fuel could not compete with that produced by the oil companies.

Interest in biodiesel was revived in the 1980s, and it was tested in buses in Malaysia using locally grown palm oil. It performed just as well as conventional diesel. In the 1990s the buses serving Logan airport in Boston, Massachusetts, ran on biodiesel made from the waste oil of restaurants in the city. The UK now has a biodiesel plant which opened in 2004 producing 16 million litres per year from waste cooking oil. However, to meet EU directives that all motor fuels must contain 8% biofuel by 2020, the UK will need to be manufacturing 330 million litres of biodiesel a year.

The global production of vegetable oils is around 120 million tonnes, of which most is still used as food and only 6 million tonnes goes into biofuel. World demand for conventional diesel is in excess of 500 million tonnes a year, so replacing this is not going to be easy. Biodiesel returns 93% more energy than is used to produce it, and unlike ordinary diesel it emits less in the way of noxious fumes, sooty particles or sulfur dioxide.

Biodiesel is derived from fats and oils. All living things produce triglycerides and these molecules consist of **fatty acids** attached to a **glycerol** molecule. The fatty acids are the biodiesel component and these have to be released from the glycerol. Converting plant oils to biodiesel is a two-step process, the first of which is to extract the oil by pressing the oilseeds, and the second step is to react it at 50 °C with sodium methoxide to form the methyl ester derivatives of the fatty acids, which is what biodiesel is.

Rape methyl ester is the usual form of biodiesel in Europe. Rape's scientific name is *Brassica napus*, but its common name comes from *rapum*, the Latin for turnip. The present yield of rapeseed oil is 3.3 tonnes (3600 litres) per hectare but genetically modified plants could double this and the target is to increase this to 5 tonnes per hectare by 2040.

Other plants whose seeds are rich sources of biodiesel are sunflower, palm and coconut. A lot of effort has gone into planting palm, whose trees take five years to become established but thereafter increase their yields to reach peak production after 20 years. Some oilseed crops will grow on marginal land that is unsuitable for farming because it is too sandy, salty, rocky or dry. Jatropha (*Jatropha curcas*) is one such plant which is a drought-resistant shrub now being grown in India, the country where it was first developed and where vast areas of it are being planted. It is also being grown in the Philippines, South East Asia, Southern Africa, and Central and South America, and the plan is to have one million hectares of jatropha by 2011, increasing this by 300 000 hectares a year thereafter. The UK company D1 Oils is planting 220 000 hectares of jatropha around the world with the aim of producing 2 billion litres of jatropha oil per year.

All over the world, old chemical plants are being converted to biodiesel production. One such plant at Bromborough, Merseyside, and owned by Lubrizol, had been shut down and was due for demolition when it was purchased by D1 Oils for $5.7 million. It would have cost them $80 million to build a new plant. The facility now processes seed crops, including jatropha, and produces more than 40 million litres of biodiesel a year. In the USA, there are more than 105 former chemical plants producing biodiesel and by 2010 their output could well exceed 8 billion litres.

Biodiesel need not only come from plant oils. Animal fats, such as tallow, chicken fat and fish oils could also be used to make it. A chicken-processing plant in the USA produces 1 million tonnes of chicken fat a year and a biodiesel plant is being constructed to convert it to 1 billion litres of biodiesel. In New Zealand it has been suggested that the tallow from sheep could be used and 1kg of this animal fat will yield around 1 litre of biodiesel.

While plant oils will be the major source of biodiesel, with animal fats contributing only a small percentage, there is another living species that might one day provide a significant amount, and these are algae. Growing in warm ponds, these tiny creatures can yield respectable amounts of oil and such ponds could use sources of low grade heat, such as waste heat from power stations. As long ago as the mid-1980s, research into alternative forms of energy at the US's Solar Energy Research Institute (SERI) showed that the algae *Chaetoceros* and *Navicula*, and the green alga *Monoraphidium*, produce a lot of oil; indeed half their weight can be oil. The algae grow by extracting carbon dioxide from the water and this is continually replaced from carbon dioxide in the air. Algae can double their numbers five times a day and they will also grow in sea water. In terms of using solar energy they are almost five times as efficient as plants. Algae don't compete with food crops when it comes to land use and could even be sited in semi-desert regions assuming there was enough water available.

Biodiesel from algae has so far produced very little, and indeed the US Department of Energy abandoned this route to a sustainable fuel future in 1996 because it cost too much. Nevertheless, there are several start-up companies basing themselves on ways of producing fuel this way, some with claims that as much as 20 000 litres of oil per hectare could be produced.

Algal ponds are not an entirely a win-win way of generating biodiesel because there has to be careful control of the system. Temperature, acidity, evaporation, salinity and invasion by alien organisms have all to be monitored, but these are not insurmountable obstacles. For example, Green Star Products of California has developed an algal strain ZX-13, which can survive temperatures as high as 45 °C and they can reproduce themselves up to four times a day given optimum conditions.[9]

In theory an algal pond could yield 100 000 litres of fuel oil per year. Biodiesel generated from algae grown on sewage waste water ponds was produced in New Zealand in 2006, by the Aquaflow Bionomic Corporation of Marlborough. Meanwhile the oil giant Shell is funding a project in Hawaii, along with HR Biopetroleum, to construct algal ponds at a site on the Kona coast. Shell claims that algae can produce 15 times more oil per hectare than land planted with rape, palm, soya or jatropha. The company PetroSun aims to produce 16 million litres of

[9] This means that one alga will become one million algae in 15 days.

algal oil from saltwater ponds in Rio Hondo, Texas, on the Gulf of Mexico.

Exxon Mobil has entered into joint partnership with Venter's biotechnology company, Synthetic Genomics, to produce biofuel from algae designed to secrete hydrocarbons that are similar to those derived from the distillation of fossil fuels.

A fungus, *Gliocladium roseum*, found in the Patagonian rainforest can convert cellulose to biodiesel, or at least into a hydrocarbon liquid.[10] It shows that a gene exists that can effect this transformation and maybe later this century a process based on this fungus will be in production.

Oils and fats can be reacted with hydrogen gas to form a different kind of biodiesel which is more like normal diesel or paraffin. This kind of diesel is sometimes referred to as *renewable* diesel to distinguish it from normal biodiesel. The process converts the fatty acid chains to hydrocarbon chains, and the glycerol ends up as propane gas. The process is being developed by Neste Oil of Finland and they have built a plant to produce 170 000 tonnes of biodiesel this way and another plant is being built in Singapore which will process 200 000 tonnes of palm kernel oil. In Australia at the Bulwer Island refinery near Brisbane, the tallow rendered from beef and sheep fat will be the raw material in a reaction which consists of mixing the fat with ordinary diesel, and then subjecting the mixture to hydrogen gas at high temperatures and pressures, using catalysts.

As a transport fuel, biodiesel is not without its problems. Its production requires a large input of energy to prepare the land; plant the crop; fertilise it; protect it with pesticides; irrigate it if necessary; harvest it; process it into biodiesel; and transport it to where it is needed. Another disadvantage is with the fuel itself. Biodiesel tends to form a jellylike paste in cold weather. Biodiesel will gel at −10 °C, although blended with ordinary diesel the gel temperature is lowered. Biodiesel produced from animal fats will even gel at temperatures as high as 15 °C. One way round this problem is for a vehicle to use ordinary diesel to start the engine, heat from which can then be used to warm the biodiesel to above its gel point.

Biodiesel can present problems for the current pipeline distribution network. The oxygen atoms of its molecules tend to cling to the walls of the pipeline and thereby contaminate the next fuel as it passes through. Around 30 million tonnes of fuel are moved by pipeline from terminals to refineries to distribution centres and were all this to be moved by

---

[10] This contains octane, 1-octene and hexadecane.

road it would require about a million tanker journeys. The fuel most sensitive to being contaminated with biodiesel is aviation fuel (jet fuel), which must not contain more than 5 ppm. Analysis for biodiesel traces requires modified GC-MS to detect its presence and measure the amount, but such analysis developed by BP is now being used across the industry.

4.3 BIOBUTANOL

What helped save the British Empire might help save the planet.

Butanol becomes economic as a transport fuel when the price of oil exceeds $40 per **barrel**. It can be added to petrol up to around 60% without the need to modify engines. Biobutanol was once produced on a large scale, not to be used as a transport fuel but as a resource for the chemical industry. A process for generating it was developed in World War I, when the British Empire was in dire need of acetone for the manufacture of cordite, the smokeless explosive for battleship ammunition.

The shortage was solved by Chaim Weizmann who worked in the Chemistry Department of Manchester University, and who had become an authority on fermentation.[11] He had shown in 1912 that it was possible to utilise the bacterium *Clostridium acetobutylicum* to produce acetone from starch on an industrial scale, and there was an ideal biomass resource in the UK from which to produce it: horse chestnuts. These are rich in starch but inedible, so vast quantities of these were collected all over the country and sent for processing at the Royal Navy Cordite Factory at Holton Heath, in Dorset.[12]

The method Chaim had discovered converted starch into a mixture of acetone, butanol and ethanol, and it became known as the ABE process. It was in operation in the US until the early 1960s because of the butanol it produced, but then it became cheaper to produce butanol by a petrochemical route. Global consumption of butanol is currently 3.5 million tonnes per year and it is used in solvents, paints and resins, as well as being a chemical feedstock in its own right.

[11] In the 1960s I did my PhD in chemistry at Manchester University in the old laboratories in which Weizmann worked and there was even a drying oven still in operation that was reputed to be one he had used.

[12] A grateful nation asked Weizmann what he would like as a reward. He requested that a state of Israel be created in British-ruled Palestine, and this led to the Balfour Declaration of 1917 which acknowledged that a homeland for the Jewish people should be established.

Biobutanol could have a future as a biofuel and it could be produced from agricultural feedstocks such as maize, wheat, sugar beet, sorghum, cassava and sugar cane. One day it might be possible to convert the non-food components of these crops to second-generation biobutanol, but there is a problem because butanol inhibits the microbes producing it, such as *Clostridium acetobutylicum*, when concentrations exceed 2%. The UK company Green Biologics of Oxfordshire, UK, have now developed a bacterium which can tolerate 4% and can digest hemi-cellulose as well. The Swiss company Butalco has modified yeasts to produce butanol instead of ethanol, and yeasts are less inhibited by butanol.

Unlike methanol and ethanol, of which there can only be one molecular form, butanol can exist in five different conformations – see Glossary for more details. That produced by the Weizmann bacterium is called the normal butanol. The bacterium *Escherichia coli* has been re-engineered to produce a different isomer, isobutanol, by a group at the University of California, Los Angeles. This form of biobutanol has a higher octane rating than normal butanol.

4.4 BIOGASOLINE (aka BIOPETROL)

The fuel that didn't save the Nazi Empire might help to save the planet.

Although much research today is directed at converting biomass into bioethanol or biobutanol, another option is to convert it to **syngas** (a mixture of hydrogen and carbon monoxide) and thence to bio-methanol, bioethanol and biogasoline, all of which are transport fuels. In fact syngas has been a source of petrol for around 80 years and was originally used to convert coal to oil. Syngas was developed by the Lurgi Company of Germany and it was converted to oil by the **Fischer–Tropsch** process patented in 1925. Franz Fischer was director of the Kaiser Wilhelm Institute for Coal Research in Mülheim an der Ruhr, where Hans Tropsch was research leader. Nazi Germany invested heavily in the Fischer–Tropsch process, and by 1944 there were more than 20 plants in operation making around 20 million litres of fuel per day. A targeted bombing campaign directed against these plants during the summer of that year crippled the Nazi war machine and helped to end World War II in Europe.

South Africa still employs Fischer–Tropsch plants to meet its energy needs and Sasol, the company operating them, turns 30 million tonnes of coal a year into transport fuels, including low-sulfur diesel which is exported to Europe. New Fischer–Tropsch plants again began to be built in the years following the Oil Crisis of 1973, and interest continues, now with the aim of using biomass as the resource. In 2007 one was opened in Freiberg, Germany, using 15 000 tonnes of biomass per year. The Finish paper producer UPM is building a Fischer–Tropsch plant to turn its waste biomass into biogasoline. Even industrial waste and domestic waste can form syngas when heated with steam. A plasma-enhanced melter (PEM) has been developed in the USA which produces syngas with high levels of hydrogen and even producing energy well in excess of that needed to power the PEM itself.

Another way of turning syngas into gasoline is the Mobil process which produces hydrocarbons *via* an artificial zeolite ZSM-5. The first large-scale Mobil plant was built in the 1980s at Motunui, on the west coast of the North Island of New Zealand.[13] This used a fossil resource, natural gas, as its raw material. No doubt it could, with some modification, use bio-syngas as well.

Fast pyrolysis of biomass, in which it is heated in the absence of oxygen, produces a liquid oil that can then be refined to syngas, and a pilot plant is being built in Germany which will produce 13 500 tonnes of this oil from wood waste and straw. Syngas can also be a source of bioethanol. A plant is being built at Edmonton, Canada, which will produce 36 million litres this way, the syngas being made from industrial waste. An American company called Coskata makes bioethanol from almost any kind of carbon-based material, and it does this by converting it to syngas and feeding this into a bioreactor, where microorganisms consume the carbon monoxide and hydrogen gas and excrete ethanol as their waste product. The process is reputed to generate seven times more energy than it consumes. The chemical company Ineos has a patented process for making bioethanol from municipal waste *via* syngas made by superheating the waste. The bacteria *Clostridium ljungdahlii* can ferment syngas at room temperature to create almost pure bioethanol and it does this quite rapidly.

One great untapped source of carbon monoxide is steel mills. Half a tonne of CO is emitted for each tonne of steel produced. Combined with hydrogen as in syngas, this could be converted to biofuels and based on world steel production it could provide more than 200 billion litres per year. How this would be done remains to be seen.

[13] This became uneconomic and was closed in 2004, although it is now being re-commissioned.

4.5 BIOMETHANOL

This fuel can be made from almost any organic waste.

Biomass converted to syngas can easily produce biomethanol and this liquid has been seen as a fuel in its own right, with certain advantages and disadvantages. Given the right conditions, the process even gives off heat.[14] Currently the world makes 33 million tonnes of methanol a year from fossil natural gas and uses it mainly as a feedstock for the plastics industry.

Methanol as M85 (85% methanol; 15% gasoline) was tested as a pollution-reducing fuel in the US in the 1980s. More than 10 000 flexi-fuel vehicles ran on M85, but the project floundered in the early 1990s when the industry turned to other ways of reducing air pollution and all research on M85 ceased in 2005. Before then it had been the preferred fuel for the Indy Racing League (IRL) and used as such in the world famous Indy 500 event, and the reason was that if there was a serious accident and a fuel tank broke open, then M85 would not create a fireball. Eventually the Indy 500 organisers plumped for bioethanol when this was offered free as part of the bioethanol industry's publicity campaign. China, however, continues to develop methanol as a fuel and around 7 billion litres are used in that country for transport.

Nobel Laureate chemist George Olah is a fervent advocate of methanol and even talks of a methanol-based economy. He has written a book advocating this chemical as the resource of the future: *Beyond Oil and Gas: The Methanol Economy*. Methanol is a non-corrosive liquid (boiling point 65 °C) which is easily transported and stored, and it burns with a clean flame producing only carbon dioxide (CO_2) and water (H_2O). It does indeed make a useful transport fuel as M85. The big disadvantage of methanol is its toxicity; it causes blindness – although only if it is drunk.

Methanol could be made from the **glycerol** by-product of biodiesel manufacture. For every 10 litres of biodiesel produced, there is 1 litre of glycerol. The US produced 2.65 billion litres of biodiesel in 2008 leading to 265 million litres of glycerol. Some is sold as animal feed and around 350 000 tonnes are burnt as fuel, but it could be converted to bio-syngas as a first step to making methanol. This is what a new Dutch company BioMCN is doing at its facility at Delfzijl, in The Netherlands, where it will convert 250 000 tonnes of glycerol into 200 000 tonnes of methanol a year. The process involves turning the glycerol firstly into bio-syngas

[14] The chemical reaction is $CO + 2H_2 \rightarrow CH_3OH$.

and thence into biomethanol. There are also plans to increase output eventually to 800 000 tonnes per year.

Biomethanol can also be produced directly from glycerol by reaction with hydrogen at 100 °C and under a pressure of 20 atmospheres in a process developed at Oxford University, and patented by a technology transfer group Isis Innovation in 2008.

4.6 BIOMETHANE (BIOGAS)

Microbes will happily produce this on a vast scale but can it run cars?

Natural gas consists mainly of methane, and compressed natural gas (CNG) is how it is used to power vehicles where it is stored in cylinders at 200 times atmospheric pressure. If these cylinders are made of a heavy metal, such as steel, it raises fuel consumption although some are now make of a lighter composite fibre/plastic material. CNG is cheap and is used to fuel cars in Canada, Europe and South America, but there are relatively few refuelling stations.

When natural gas supplies run out, then biomethane could be used. This is produced when **anaerobic** microbes digest organic matter. Methane can comprise as much as 75% of biogas with carbon dioxide accounting for the rest. It is released in vast amounts from old landfill sites and is being gathered from these sources and burnt to generate electricity. In Canada about 25% of such sites are tapped for their biogas and produce 100 MW of electricity. The Capstone Turbine Corporation in California manufactures 30-kilowatt and 60-kilowatt micro turbines for this purpose, and more than 2000 of these have now been sold worldwide. The turbines burn up to 250 cubic metres of biogas a day.

A cubic metre of biogas will provide as much energy as 0.75 litre of petrol, so it has potential as a transport fuel. There is a site near Albury in Surrey, England, which contains an estimated six million tonnes of municipal waste. Its biogas is collected and compressed for use in vehicles. It is referred to as liquid biomethane, or LBM, and the site produces around 5000 tonnes of LBM per year, enough to fuel hundreds of heavy goods vehicles.

Almost any organic material which can be reduced to a slurry can be fermented rapidly with anaerobic bacteria and the biogas collected. Bananas that have been rejected as unfit for sale in Queensland, Australia, are used to make biogas. Seaweed can be a source, generated by a process developed by the Scottish Association for Marine Science. Rice straw, of which China produces 230 million tonnes a year, is another resource, and to make it more easily biodegradable this is ground up with a little sodium hydroxide before being steeped in water to produce increased amounts of biogas.

Biogas consists not only of methane and carbon dioxide, it also contains hydrogen sulfide, a highly reactive and toxic gas. This means that biodigesters have to be made of resistant metals, like stainless steel, and the hydrogen sulfide has to be removed before the gas can be used as a transport fuel. Perhaps not surprisingly, it is much easier to burn biogas to generate electricity than to clean it up to make it acceptable as a transport fuel, and it is likely that this will always be the preferred option.

4.7 BIOHYDROGEN

Is the much hyped hydrogen gas economy mainly just hot air?

Gram-for-gram, hydrogen packs more punch in energy terms than any of the other fuels discussed in this chapter, but it has problems which may defeat the hopes of those who foresee a hydrogen-based economy. On the face of it, hydrogen looks attractive but there is a slight snag: it is a very light gas, which means that in *volume* terms it stores only an eighth of the energy of gasoline, and that's even when it's compressed to 200 atmospheres. Of course, as a liquid its volume is much smaller, but much, much colder, *minus* 253 °C in fact. In theory hydrogen could fuel all forms of transport, including aircraft. On land it could be used in fuel cells to run vehicles powered by electric motors, but few such vehicles exist.

As we have seen, syngas processes produce vast quantities of hydrogen and this can be separated out with the aid of a silver–palladium membrane, which allows hydrogen gas to pass through but not the carbon monoxide. The hydrogen so obtained is 99.99% pure. A quicker, if less effective, method of removing the carbon monoxide is with a zeolite which can trap larger gases but not hydrogen. At the present time hydrogen is also obtained as a by-product of chlorine production, and chemical industries around the world produce 40 million tonnes a year of it. For a hydrogen-fuelled economy we would need at least 20 times as much. Transporting it in bulk as a liquid will not be a problem because there already are tankers which can carry 75 000 litres (5 tonnes). However, there are other ways of moving it around, as we shall see.

Hydrogen as a transport fuel has its challenges, some of which were overcome in the 1960s because hydrogen was vital to the US space programme. Storing liquid hydrogen was important and a tank at Cape

Figure 4.2 Some cars already run on hydrogen made from bread and water.

Kennedy can hold more than 3.5 million litres, and which loses only 0.03% of its capacity per day.[15] A Saturn V rocket took 32 hours to fuel-up and required 92 loads each of 25 000 litres of liquid hydrogen. Clearly the problems of producing hydrogen, transporting it, storing it and handling it had been mastered.

Carrying hydrogen for use in a motor car presents a different kind of problem. Ideally its storage should not require more space than currently occupied by the fuel tank. The volume of gas needed to travel 400 km would fill a balloon 5 metres in diameter and while this volume could be compressed into a cylinder, it would be heavy and bulky. Vehicles could carry hydrogen as liquid hydrogen, and in Hamburg there is a filling station that supplies this for six specially adapted vans, and at Munich airport there is also a bus which runs on it. There are even liquid hydrogen refuelling stations in Singapore, Berlin, Sindelfingen (near Stuttgart) and Sacramento, California. The BMW liquid hydrogen car holds 120 litres of the fuel in a tank whose insulation consists of 70 thin layers of aluminium and glass fibre, vacuum-packed into a 3 cm gap between two metal skins. Refuelling takes only 3 minutes and is done automatically with no loss of hydrogen.

[15] Which doesn't sound a lot but this is equivalent to 10% per year.

Californian Governor Arnold Schwarzenegger issued an executive order in 2004 to the effect that hundreds of hydrogen-fuelling stations would be built in that state, but not a single one has been constructed. Mary Nichols of the state Air Resources Board is still optimistic that 50 to 100 stations will be built by 2015 but there are fewer than 200 hydrogen vehicles on Californian roads.

Syngas is one way of producing hydrogen but there are other ways. We can do it easily, but expensively, with electrical energy, or cheaply, but with difficulty, by using light energy from the sun. It has been known for more than 30 years that in the presence of **titanium dioxide**, ultraviolet light will decompose water into its constituent elements. To date yields of the gas are tiny and represent less than 1% energy conversion. In 2002 a group of scientists, led by Shahed Khan at Duquesne University in Pittsburgh, Pennsylvania, boosted the efficiency of titanium dioxide by burning the metal at 850 °C. This activated titanium dioxide releases much more hydrogen and with an efficiency of energy conversion of 8%.

Nature has evolved its own hydrogen-generating enzymes, the *hydrogenases*, and these offer a milder way of producing the gas from organic matter, and indeed it is produced in small amounts in our own intestines in this way. When conditions are just right, species like green algae and cyanobacteria will emit hydrogen. In 2004 a group of Japanese chemists, led by Professor Naomichi Nishio of Hiroshima University, discovered a new bacterium which can digest waste from the food industries, such as bread, and the spent liquid from brewing soy sauce and rice wine, and release hydrogen gas. One day it may be possible to ferment organic wastes with genetically engineered bacteria and release hydrogen in useable amounts. That day is probably still far into the future and the process will undoubtedly be slow, because the enzymes that generate the gas have their active site buried deep within the enzyme to protect them against oxygen.

Using sunlight and a metal catalyst to generate hydrogen from water has relied up to now on the UV part of the spectrum, and even then was inefficient. Now a group at the Dalian Institute of Chemical Physics, in China, have found a catalyst which can produce hydrogen from water using **visible light** in the 420 nm (violet) region and with a quantum efficiency of more than 90%. The new catalyst was made from cadmium sulfide doped with palladium sulfide and platinum.

Chemists have devised ingenious ways of storing hydrogen, the best of which is to absorb it into another material, and there are some that are particularly good at doing this. Alloys such as iron–titanium and lanthanum–nickel can absorb so much hydrogen it is as if it were the same amount of liquid hydrogen. A van that ran on hydrogen gas stored in such an alloy was demonstrated in 1986. The gas released from the alloy was pumped into the engine and when the engine had warmed up, the steam emitted from it was used to heat the alloy, releasing more hydrogen. However, that alloy slowly lost its hydrogen-absorbing ability. This problem has been solved with an alloy of magnesium, nickel and aluminium, which has been patented by the US-based company Ergenic and is called Hy-Stor. It can be charged and discharged more than 100 000 times.

A promising substance in which hydrogen might be stored is one that consists of nanocubes, and these are already being produced by the chemical company BASF. They were discovered by Omar Yaghi of the University of Michigan in 1999. The material has open lattices with cube-like voids and is made by reacting zinc oxide with special acids. The empty cavities in this compound are so large that the material itself has the lowest density of any crystalline solid ever recorded at $0.21\,g\,cm^{-3}$.[16] Nor have these voids to be packed with molecules to support this open framework, the material retains its shape even when its pores are empty, and it is robust enough to be heated to 400 °C.

Hydrogen can simply be burned as a gas to release the energy to transport us but it is much better to use it in a **fuel cell**. The world's first bus driven by such fuel cells was exhibited in Long Beach, California, in 1994. It had a top speed of 74 kph mph and a range of 160 miles before it needed refuelling. In a fuel cell, hydrogen still reacts with the oxygen of the air, but it does so in a controlled way and giving up most of its energy to generate electricity, rather than as heat, although some heat is produced. The British scientist William Grove was the first to demonstrate a fuel cell which he did as long ago as 1839. His devices were known as Grove cells. However, it was not until 1956 that Francis Thomas Bacon, an engineer at Cambridge University, England, showed that it could be done in a practical manner. (In the 1960s, fuel cells were used in the Gemini and Apollo space programmes.)

The favoured fuel cell for automobiles is the proton electrolyte membrane, and a stack of these comprising a 75 kW power unit takes up about the same space as a conventional car engine. The motor manufacturer DaimlerChrysler has produced a fuel-cell powered bus called the Citaro, in which the hydrogen and the fuel cells are held in the roof. The Honda motor company began making vehicles with hydrogen fuel cell technology in 2003, and these are available in the USA, although

[16] Zinc oxide itself has a density of $5.6\,g\,cm^{-3}$ which is 25 times heavier.

refuelling them is somewhat of a problem because of the lack of refuelling stations. Hybrid cars are also being developed and these run on both a fuel cell and a small internal combustion engine, which kicks in when the vehicle is running at a constant speed. Such hybrid cars achieve an overall efficiency of 60%, double that of a normal car. What restricts their use is their need for platinum. A typical fuel cell requires about 50 grams of this rare metal which adds $2000 to the price of a car.

One must conclude that hydrogen gas as a transport fuel for the family car is not really going to happen, and in any case it will first have to shed its popular image of being a particularly dangerous fuel. The dramatic film of what befell the hydrogen-filled Hindenburg airship in 1937 has coloured people's view of what hydrogen is capable of doing.

Before we close this chapter there is a simple question to consider. Would it be possible to produce all transport fuel from sustainable sources? In an ideal world we would grow crops for food, turn their inedible carbon-based components to biofuel, and return their non-carbon nutrients to the soil as fertiliser. In an ideal world.

In developed economies there are around 50 cars per 100 people, suggesting we might end up with a total of 4.5 billion cars when there are the expected 9 billion humans on the planet and if all were to enjoy the same standard of living. The thought of 4.5 billion cars is frightening, but what depresses a scientist like me is that they might all be as inefficient in their use of fuel as cars of today. A car wastes far more fuel than it needs merely to transport someone from A to B. Hybrid cars are less wasteful and some can travel more than twice as far on a litre of fuel than current motors. Even so, what a car is mostly using its fuel for is to produce heat and move its own weight.

In 2008, London's Royal Society issued a report: *Sustainable Biofuels: Prospects and Challenges*, which said that renewable fuel policies were encouraging the wrong type of biofuels which don't cut greenhouse gases and could cause social and environmental problems. The Netherlands Environmental Assessment Agency says it would be better to burn biomass to make electricity than to convert it to transport fuels. Of course no matter what these prestigious bodies say, one day the main source of accessible transport fuels will have to be the biomass or hydrogen from water.

Sustainable transport fuels such as bioethanol and biodiesel were rather over hyped by environmentalists in the 1990s and early 2000s, and few questioned that they would be anything other than a beneficial substitute for fossil fuels. Today we realise that the cost of sustainable fuels are also subject to the laws of supply and demand, and despite tax incentives and legal requirements to add them to existing fuels, their future now seems less assured. In 2008 many biofuel plants in the EU were operating at a loss because even though demand for bioethanol was running at around 2 billion litres a year, there was capacity to produce more than 3 billion litres. In Germany where more than 3 billion litres of biodiesel are produced from rapeseed oil, the cost of producing it was becoming prohibitive and yet there is capacity to manufacture 5 billion litres. The economics of biodiesel production was also skewed by the US where biodiesel blended with ordinary diesel qualifies for a tax credit of $1 per US gallon (3.8 litres), which has boosted imports from South America and Asia. More than 500 million litres of this blended biodiesel was then being re-exported to the EU, where it is cheaper than locally produced biodiesel. Again this is not a problem which chemistry can solve.

Chemists can turn almost any form of carbon-containing biomass into transport fuels, but it would be unrealistic to provide biofuels for a world with billions of inefficient private motorcars. Solve that problem and a sustainable future for most forms of transport, including large numbers of private cars, might just be possible. However, it seems unlikely that the fuel demands of air transport can also be met from carbon-based biofuels. Hydrogen may well be the jet fuel of the future.

CHAPTER 5

Plastics and Chemistry

[A word in **bold** means there is more information in the Glossary.]

There are seven plastic recycling codes, of which numbers 1 to 6 are for the most common varieties, while 7 covers the rest (see page 104). Could they all be made from sustainable resources or maybe replaced with bioplastics? The seven topics in this chapter are: biopolymers, polyethylene and polypropylene, PVC, polyester, polystyrene, polyurethane and extreme polymers.

In 1988 I wrote an article for the popular science magazine *New Scientist* called 'Plant a tree for chemistry', in which I explained to readers how these plants were a resource for the chemical industry and how they might well furnish even more materials in the future when oil supplies ran out. That time has yet to come, but interest in biomass, of which trees constitute only part, as a source of sustainable plastics has become a reality, as yet contributing only a small proportion of the plastics we rely on.

Everyone else calls them plastics; chemists call them **polymers**. But whatever their name, they are the materials which have moulded the modern world. More than 300 million tonnes are manufactured every year and almost all are made from oil. One day they will have to be made from renewable chemicals – or replaced with biopolymers, or even reclaimed from old landfill sites, or we will have to do without them. Such are their benefits, that this last possibility is not really an option in a developed society, and is not likely to happen unless there is a lack of investment in green chemistry and young chemists.

A Healthy, Wealthy, Sustainable World
By John Emsley

Published by the Royal Society of Chemistry, www.rsc.org

Figure 5.1 Plant a tree for chemistry.

We should ask of polymers – indeed of all materials – that they pass the three Rs test of sustainability which is: reduce, reuse, recycle. Reducing their use is already happening as in the case of supermarket plastic bags which are no longer handed out freely and whose numbers have reduced accordingly. Reuse is possible for some plastics, such as plastic bottles, but few countries adopt this approach. Recycling polymers is a much better option because they can easily be turned into new products, provided they are collected and separated into their various types.

A lot of recycling is done by hand but at some waste facilities it is done automatically. Individual plastics can be identified and separated using infrared scanners. When it comes to recycling plastics, some fare better than others. In the USA while 620 000 tonnes of PET and 475 000 tonnes of HDPE are recycled, only 60 000 of polypropylene ends up this way. However, it is not the particular kind of plastic which determines whether it is recycled – in theory all can be recycled – it's the kind of product for which it is used that is the key factor. What a recycling facility needs is a steady supply of one particular plastic and that requires a consistent policy of collection and sorting. Unfortunately this depends very much on local authorities and their approach to waste

Common Sense 12: The trouble with most plastics is that they are either not biodegradable or they can't be recycled

Wrong. Biodegradable plastics were once thought to be the answer to plastic waste and some were developed and manufactured, but this approach is no longer seen as the way forward. Instead, plastics should be recycled and some already are, but they may re-emerge in different guises. For example, plastic bottles are turned into clothes and carpets. Plastics now carry a recycling logo with a number to indicate which type of plastic it is: 1 indicates polyester (PET); 2 is high-density polyethylene (HDPE); 3 is polyvinyl chloride (**PVC**); 4 is low-density polyethylene (LDPE); 5 is polypropylene (PP); 6 is **polystyrene** (PS); and 7 covers all other plastics, such as **polyurethane** and polycarbonate. Many polymers can be melted down and reused or turned into other products, or even reduced to the raw materials from which they were originally made.

Plastic waste amounts to 25 million tonnes in Europe annually and increases year by year. Much of this could be recycled were it to be separated and, even when this is not possible, there are plans to shred it and convert it to plastic boards which could replace the plywood boarding used on a large scale by the construction industry, and which is discarded after being used only once.

management, and in the UK it varies from locality to locality. Some day there will have to be a national policy on recycling, but until this occurs a lot of potentially recyclable plastic will end up as landfill.

"Life is plastic, it's fantastic" sang Aqua in the 1989 pop song *Barbie Girl*, and indeed they are fantastic, although the song was mocking them. Back then a lot of plastic ended up as household waste, and non-biodegradable waste at that. The answer appeared to be 'bio' polymers, meaning ones which biodegraded easily. However, the word 'bio' in the context of polymers can mean one of two things: it can mean a material which biodegrades in the same way that plant debris is broken down, or it can mean a polymer made from biologically derived materials. Because this chapter is about sustainability it is the latter we shall be concentrating on.

In the past few years, plastics have regained some of their former glory as wonderful materials which solve more problems than they create. The challenge now is to make them sustainable. There are three ways this could be done: find alternative sources of raw material to make existing kinds of plastic; find alternative plastics which can be made from sustainable resources; or revert to using natural plastics, the kind of biopolymers which humans have used for thousands of years, and which Nature produces on a vast scale, such as **polysaccharide** carbohydrates and polypeptide proteins. These are things like wood, wool, cotton, silk and horn. Rubber is a natural hydrocarbon biopolymer made from the sap of the rubber tree *Hevea brasiliensis.*[1] Clearly biopolymers are renewable but they have their failings. The attractions of the man-made polymers discovered in the last century were that they were more versatile, were cheaper, more colourful, stronger and safer.

In Europe, bioplastics account for a mere 0.1% of the polymer market: 50 000 tonnes compared to 50 million tonnes of oil-based ones. Clearly we have a long way to go to replace the remaining 99.9% sustainably. In this century chemists need to find ways to do this or discovery bioplastic alternatives.

Plastics in the EU are reputed to save more then 22 million tonnes of crude oil per year, and they do this by making products which last longer; by reducing transport costs in terms of lighter vehicles, including aircraft; by reducing the gross tonnage of goods being transported; and by making effective insulation. Today plastics constitute around 10% of the weight of a car, and this is increasing, and there is significantly more in aircraft. The new Airbus A380 is 25% plastic, while the Boeing 787 Dreamliner will be nearer 50%. The former will be the first long-haul carrier to consume less than 3 litres of fuel per passenger per 100 km of flight.[2]

5.1 BIOPOLYMERS

Chemists have been looking for sustainable plastics for many years but the ones they come up with are not yet economically competitive.

In 2008 the production of biopolymers was mainly natural rubber (9 million tonnes) and **rayon** (4 million tonnes). Other biopolymers accounted for only 500 000 tonnes, and much of this was just one bioplastic, polylactic acid. This is one of the new generation of biopolymers, in this case produced by microbes. Another biopolymer that promises to have a future is polyhydroxyalkanoate, and the US company Metabolix has a strain of the bacteria *Escherichia coli* which is capable of forming it. This can be formulated to compete with

[1] This contains the simple hydrocarbon 1,4-isoprene which polymerises when heated.

[2] This double-decker aircraft will carry 525 passengers, so that the 5570 km flight from London to New York will consume 70 tonnes of fuel.

man-made polymers like polyethylene and making it does not require much energy, but as yet it cannot compete in price.

Another biopolymer is **polyamide**-11 (aka nylon-11), which is a natural polymer that has been around since the 1930s and it is competitive. It is made by polymerising the naturally-sourced chemical 11-amino-undecanoic acid which comes from castor oil. The reason it has not been replaced by man-made **nylon** is that it has some unique properties which make it ideal for gas distribution pipelines and air-brake hoses,[3] in that it does not swell when exposed to water and it is not affected by oils and grease. Around 150 000 tonnes of polyamide-11 are made every year.

Plants produce one stable biopolymer superbly well, and that is cellulose. This is made from **glucose**, which in turn is made from water and from CO_2 taken from the air. Glucose is the most abundant biological molecule produced on Earth, to the extent of around 200 billion tons a year, equivalent to 80 billion tonnes of carbon[4] and more than ten times the amount of fossil carbon humans extract from the planet's crust.

Glucose can link itself together in long chains of two kinds. The first is one which can be easily broken down again into its constituent units by enzymes, and we know this as starch. The second is one which is almost indestructible and can last for thousands of years, and we know this as cellulose. A plant uses starch to store energy, while it uses cellulose to constructs its roots, stems and leaves. The difference between starch and cellulose is simply in the way in which the glucose molecules are joined to one another.

Natural cellulose can come in the form of a fibre ideal for making fabrics, such as cotton and linen, but a lot of cellulose comes in less useful forms, such as wood. The cellulose can be extracted from such materials, however, and turned into fibre, such as rayon, or into the transparent film known as **cellophane**. To extract and convert cellulose into fibres for textiles it is necessary to solubilise it. This is not easy because the chains of cellulose polymer cling strongly together. The only way to untangle them and pull them out into fine threads is to dissolve them. For rayon this is done with caustic soda solution and carbon disulfide. The process is efficient but produces a lot of foul-smelling effluent. When the viscous solution that ensues is forced through tiny nozzles it forms fibres of rayon, or through a narrow slit it emerges as

[3] Air-brakes are used in trains and rely on air pressure to keep the brakes open. When the pressure drops the brakes are applied.
[4] Glucose is $C_6H_{12}O_6$ with a formula weight of 180, of which its 6 carbon constitute 72 parts, *i.e.* 40%.

cellophane. As it emerges, the solution is neutralised with acid to make the cellulose insoluble again.

Rayon has many advantages: it feels gentle to the skin, it hangs nicely and it dyes well. Rayon is widely used for underclothes, blouses, dresses, jacket linings, and blended with other fibres in bed linen, upholstery and curtains. But it creases easily and lacks strength when wet.

For 80 years the chemistry of rayon manufacture changed little, until research chemists discovered that cellulose would dissolve in the solvent ***N*-methyl morpholine oxide**. This is the basis for the production of a newer form of rayon called Tencel and Lyocell. This process keeps the cellulose fibres longer and stronger, and there is little or no environmental pollution. The fibre is extruded into air and immediately passed into water to wash out the solvent which is reclaimed for reuse. Tencel not only has a more luxurious feel and fluid drape, but it also has better wet strength, lower shrinkage and is crease-resistant. It is used in bath towels, and as clothing it presents a silky appearance and used in things like dress shirts and blouses. In Australia Tencel is made from wood pulp harvested from managed eucalyptus forests which grow on marginal land unsuited to food production. These trees mature after 10 years and produce around four times as much cellulose as cotton, which generally grows on farmland and needs both irrigation and pesticides.

Natural rubber is a biopolymer formed from the latex of the tropical tree *Hevea brasiliensis*. The simple chemical **isoprene**, which this contains, easily polymerises to form an elastomer or, in other words, a stretchable polymer. This can be modified in many ways to provide a range of useful materials, from condoms to car tyres. But natural rubber is not entirely free of impurities. It has an inherent structural weakness in being susceptible to chemical reaction with oxygen which leads to its discolouration, hardening and cracking. Also its enzyme impurities can provoke an allergic reaction in some people. Natural rubber cannot meet global demands for elastomers, which are now mainly of the synthetic variety and offer an even wider range of properties, from chewing gum to quieter road surfaces.

Styrene, butadiene and isoprene are the major feedstocks used in the manufacture of synthetic rubbers, and by co-polymerising these it is possible to get this wide range of useful materials. Artificial isoprene rubber is preferred wherever a high-purity elastomer is needed – and that is why it is in great demand for medical and pharmaceutical uses. Catheters, surgical gloves and gastric balloons are just some of the products made from it. Even medical workers who are allergic to natural latex gloves can safely wear those made of synthetic rubber.

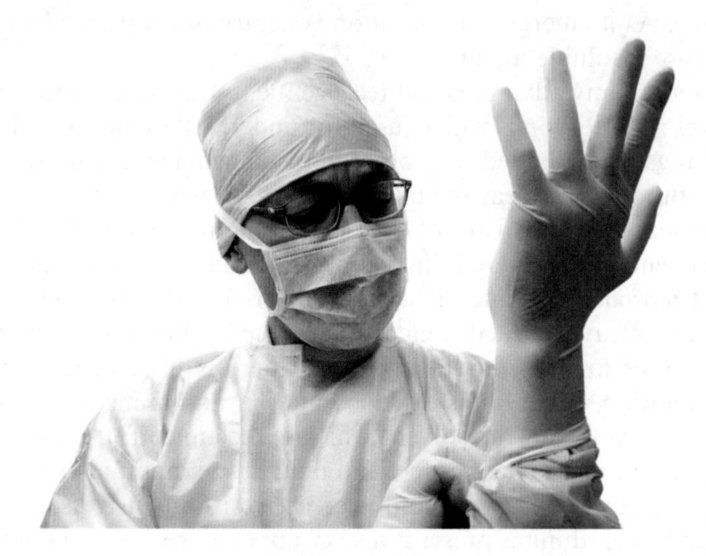

Figure 5.2 Some medics were allergic to latex gloves, but that's no longer a problem.

Polybutadiene is another synthetic rubber which is widely used in the tyre industry and also in the production of toughened polystyrene for the manufacture of packaging, toys and consumer durables. Rubber made from butadiene will still be needed in a sustainable future, so ways of making these raw materials will have to be found. A genetically engineered *Escherichia coli* has been produced which can make isobutanol (2-methylpropanol) and this can be dehydrated to produce isobutylene (2-methylpropene), a useful raw material for the chemical industry and from which polyisobutylene rubber can be made.

One man-made bioplastic was designed specifically to meet the requirements of the 1980s that polymers should be biodegradable. It was hailed as a success and was suitable for all kinds of things including films and fibres. Its story illustrates what chemists can achieve, but it shows that the dictates of society are not always the best guide.

In the 1960s and 70s polyethylene became too popular in the form of plastic bags, billions of which were handed out to shoppers at supermarkets. Some ended up adorning trees, fences and power lines, never appearing to rot away. People objected and called for biodegradable plastics so that when their usefulness was over they would quickly disintegrate. Chemists came up with a biodegradable plastic extracted from bacteria which produce it as a way of storing energy. Not only could plastic bags be made from it, but it could be turned into all kind of products. The polymer was **poly(hydroxybutyrate)** or PHB for short. Its trade name was Biopol.

PHB was first isolated in the 1920s by Maurice Lemoigne at the Pasteur Institute in Paris from the bacterium *Alicaligenes eutrophus*. In the 1970s the US chemical company W.R. Grace & Co investigated it, but it was left to the UK chemical company ICI to develop a large-scale fermentation process for manufacturing it in the 1980s. PHB had some useful properties: it did not break down in water, which other biodegradable polymers are prone to do; it was strong; and it was stable to oxygen.

PHB appeared to have great potential and it could be modified by including varying amounts of a similar acid, 3-hydroxypentanoic acid, in the fermentation broth which the bacteria would also incorporate into the polymer.[5] The more of this was included, the softer the polymer became, thereby extending its uses. And it conformed to the demands of the public with respect to its biodegradability. It would break down in sewage, soil, seawater and even in the stomachs of cattle. It was ideal for things like disposable nappies, sanitary pads and incontinence pants. It was even tested as the capsule for slow-release drugs.

Biopol's future seemed assured when it was chosen by Wella for their Sanara shampoo bottles. This was named product of the year in Germany in 1990. Soon Biopol was being used as cosmetics packaging in countries like Italy and Japan. In the US the readers of *Popular Science* magazine voted it one of 1990's greatest scientific achievements. In 1997 the Co-operative Bank in the UK, with the support of Greenpeace, began using Biopol for their credit cards in place of **PVC**. There was talk of Biopol production reaching 10 000 tonnes per year. In 1995 biologists at the Carnegie Institution of Washington, based at their Stanford, California site, generically modified thale cress (*Arabidopsis thalinana*) so that 14% of its dry weight was PHB and these plants seemed happy to produce it.

When ICI split off its pharmaceutical division as Zeneca, the BHP production became part of the new group and in 1996 that company sold the Biopol business to Monsanto, who said that it would be produced on a larger scale and used to make disposable plates, cups and cutlery. However, this was not to be because Biopol still had a serious commercial flaw: it was much more expensive than conventional polymers. In 1999 Monsanto stopped work on it and in 2001 they sold the company to Metabolix.[6] By now PHB was being made more cheaply by the microbe *Escherichia coli*. In 2008 Metabolix announced it had

[5] Such material is often referred to collectively as PHA or short polyhydroxyalkanoate.
[6] This company won the US presidential green challenge award in 2005 for its work on bioplastic technology.

genetically engineered switchgrass to produce 3.5% of the polymer within that plant's cell walls and was aiming for a yield of around 7% in order to make it worthwhile for commercial production. As we saw in a previous chapter, this plant can generate a billion tonnes a year of biomass and would not adversely affect food production because it could be grown on land unsuitable for farming. However, Biopol's *raison d'être* has now become a millstone round it neck and that is its biodegradability. This is seen as a major drawback today because its decomposition products are greenhouse gases.

There are newer biopolymers which are more successful and one such is polylactic acid (aka PLA) made from **lactic acid**. This is produced by the fermentation of carbohydrate and the raw material can be whey, dextrose, starch or molasses. PLA is suitable for packaging and textiles, and the solid form resembles polystyrene, while its transparent film is like cellophane, and its fibre is similar to polyester and suitable for making T-shirts and carpets. PLA is also used for soft furnishing fibres and for credit cards. PLA is the basis of Lactron, a fibre which produces clothing with a softer feel on the skin than cotton or polyester, and it does not affect those who are susceptible to atopic dermatitis. As a plastic, PLA can also be coloured and can incorporate a UV filter. Packing salad vegetables in PLA gives them a longer shelf life because this plastic will allow water vapour to diffuse out of the pack. The throwaway transparent beakers that you find near water coolers are most likely made from PLA. Bioplastic PLA can be composted or chemically recycled back to the feedstock lactic acid.

The leading PLA producer, making 140 000 tonnes a year, is NatureWorks, which is jointly owned by the US agricultural group Cargill and the Japanese fibre maker Teijin. Its plant is located at Blair in Nebraska, and it makes its lactic acid from maize or sugar beet. The company hopes one day to produce PLA from cellulose feedstock. Its main product is called Ingeo and is used for plastic water bottles. A similar facility exists in China operated by Tianan and in the Netherlands operated by Hycail, which produces 50 000 tonnes a year. The car giant, Toyota, is investing in a plant to make PLA.

PLA is far from perfect as a plastic and cannot be used for many applications but all this might change. Stretching a plastic like polypropylene makes for a stronger film of the type used in packaging and the same happens with PLA. It can then even be metallised by putting it under vacuum conditions and evaporating aluminium from a heated wire. The metal deposits as a mirror on the film of plastic.

Biopolymers clearly have a future but can we continue to enjoy the benefits of existing polymers, which have many remarkable properties, by making them from sustainable resources, in other words from biomass?

5.2 POLYETHYLENE (aka POLYTHENE, PE) and POLYPROPYLENE (PP)

These come in many guises, and are the most common polymers.

Polyethylene was discovered in 1933 by Reginald Gibson and Eric Fawcett at ICI in the UK. They produced what is now called low-density polyethylene (LDPE) which is made at high pressures and this is the plastic of many plastic bags, while high-density polyethylene (HDPE) is made at lower pressures by means of special catalysts and this is used to make containers and pipes. The length of the polymer chain depends very much on the temperature at which the polymerisation is carried out: the lower this is, the longer the chain, and it was catalysts which made this possible, and in particular those based on titanium and aluminium. The length of a polymer chain has a marked effect on properties, such as its softening temperature, flexibility and toughness.

Polyethylene is produced at a rate of 70 million tonnes a year from ethylene gas, which is produced from ethane, propane, **naphtha** and **gasoil**.

LDPE has applications as cable insulation, and as surface coating for paper and cardboard. HDPE has less cross-linking between the polymer chains than LDPE, and consequently its chains pack more closely giving it a higher density, hence its name, making it stronger and more rigid. HDPE is used to make products such as bottles, buckets, bowls, food boxes and pipes. A variant of LDPE known as linear low-density polyethylene (LLDPE) is made in a co-polymerisation process under conditions similar to those used for HDPE. Its low density comes from the polymerising of a mixture of ethylene with other olefins, such as butene-1, hexene-1 or octene-1, which introduce side chains into the polymer that prevent close-packing, hence the lower density. LLDPE is stronger, irritatingly so when we come to open mail wrapped in it.

Wonderful as polyethylene is, it is far from perfect. Polythene bottles can hold water but they tend to soften and even dissolve into holes when other liquids are put in them. This problem can be solved by coating the surface of polythene with a thin layer of a tougher plastic. This layer is made by exposing the polythene container to fluorine gas which reacts with the surface CH_2 units and converts them to the more stable CF_2. The fluoropolymer layer gives the container two advantages – it makes the surface resistant to all forms of chemical attack, and it repels all other liquids. The treated polythene can now hold oil, cleaning fluids, printing inks, cosmetics, toiletries and even the most concentrated sulfuric acid. Such is the strength of fluorinated polythene that it is used for fuel tanks.

Polyethylene can be made from sustainable resources. All it needs is a source of ethylene and this is easily made. The first chemical plant making biopolyethylene will begin operating in the state of Rio Grande do Sul, in Brazil in 2011. This will be the largest renewable plastics plant in the world which will convert ethanol derived from sugar cane into ethylene and thence into polyethylene. The companies involved are Ecosurf, a subsidiary of Dow, and the Brazilian company Crystalsev. Provided the price of oil remains around $50 a barrel then it will be able to make polyethylene at a price competitive with that made from oil. The new plant will produce 200 000 tonnes a year of LDPE.

Ethylene (aka ethene) is but one kind of olefin and it consists of two carbon atoms joined by a double bond. Increase the number of carbon atoms to three and you have propylene (aka propene). Propylene is also a gas, and its major end-uses are **polypropylene** and **acrylonitrile**, which is converted to acrylic fibres and plastics. Polypropylene can range from being a rigid plastic to a fluffy fibre; it can be crystal clear or multicoloured; as soft as silk or as hard as iron. Polypropylene is heat stable and sterilisable, impermeable yet flexible, transparent and perfectly safe. There are items made from polypropylene in most homes: the kettle, rugs, suitcases, thermal vests, disposable nappies and DVD cases. Cars contain a lot of polypropylene in the form of dashboards, bumpers, battery cases, upholstery and carpets. Some cars contain as much as 80 kilograms of polypropylene. Polypropylene is also widely employed for packaging medical products and appliances, such as vials and disposable syringes.

While polyethylene is produced in grades which are distinguished by their density, polypropylene is produced in ways which depend on the chemical structure of the polymer chains. Like polyethylene, polypropylene has a backbone of carbon atoms in long chains but it also has a methyl group (CH_3) attached to every other one. This gives polypropylene a versatility which polyethylene does not have.

The ability of polypropylene fibres to stretch accounts for some of its major uses as tapes and fibres. The tape is woven into sacks, carpet backing and ropes. Fibre polypropylene ends up as carpet pile, blankets, upholstery, wall covering, underwear and sports goods. There are several types of melt-spun fibre some of which are suitable as sewing thread, netting, filters and even engineering constructions. Part of polypropylene's appeal is its soft feel against the skin, and it has the benefit of allowing moisture to escape so it is used in thermal vests and even space suits for astronauts.

Polypropylene is eminently recyclable. The automobile industry is well ahead in this respect, with polypropylene bumpers and battery cases being a major source of material for recycling. Other polypropylene artefacts that are recycled are milk bottle crates and beer bottle crates.

Propylene currently comes from oil but it could also come from sustainable sources. One obvious source would be glycol, the three-carbon molecule that is a by-product of biodiesel and of which there is currently a glut on the world market, as we saw in Chapter 4. Alternatively the long-chains of biodiesel itself might be broken down into shorter chains and propylene obtained from these. Bio-polypropylene is yet to be produced.

5.3 POLYVINYL CHLORIDE (aka VINYL and PVC)

This much maligned polymer has much to recommend it.

In India there are plans to convert bioethylene into PVC, and while this will be more expensive than that made from oil it will become competitive as the price of oil increases, and one day all PVC will have to be made this way, so it is a potentially sustainable plastic.

PVC consists of long chains of carbon atoms to which are attached chlorine atoms on alternate carbon atoms. These chlorines provide toughness, increased fire resistance and give the polymer the longevity which is deemed essential for many of its uses, such as water pipes, window frames and floor coverings. Despite what some say, PVC is manufactured to a high standard that ensures it is perfectly safe to use, so much so that it is the preferred material for blood bags, for tubing used in surgery and for the 'second skin' used to treat burns victims. Nor is PVC damaging to the environment – its longevity as a plastic ensures that PVC pipes will last many times longer than other materials. PVC is the preferred plastic of electrical wiring and cabling because it offers strength and safety and it can be stripped away when no longer required and recycled.

There many factors favouring this long-lasting and fire-resistant plastic. When PVC window frames with double glazing are installed they need no maintenance for at least 25 years, as well as providing excellent heat and sound insulation. When their life is over the PVC can be melted down and re-cast into other useful items, such as garden chairs, park benches and fencing, and when these are no longer needed they can be melted down again and used to construct drainpipes, and when finally these are no longer required, the plastic could even be melted down yet again and maybe used to make road surfacing.

What worries some people is that in order to make PVC pliable – its normal state is to be strong and hard such as in window frames – it requires the addition of a plasticiser and the ones which were commonly used were phthalates. Phthalates are needed for tubing, electrical wiring, flooring, household containers – and shower curtains. These molecules act to lubricate the polymer molecules so that they can move easily with respect to one another and without causing the plastic to crack. It is said that phthalates are endocrine disrupters, *i.e.* they interfere with hormones. There are various kinds of phthalates, such as dimethyl phthalate, diethyl phthalate and diisobutyl phthalate, and some have undergone EU risk assessments. The result was that no new safety measures were deemed necessary in order to protect human health or the environment. Nevertheless, some phthalates are banned for use in toys and baby care products, and alternatives have been substituted.[7]

As its name indicates, PVC is made by polymerising **vinyl chloride**, which is made from ethylene and chlorine, so in that respect PVC could be a sustainable bioplastic if the former is produced as indicated in Section 5.2 above, and chlorine will always be sustainable so long as there is salt in the sea.

5.4 POLYESTER (aka POLYETHYLENE TEREPHTHALATE, and PET)

A popular product of the chemical industry and easily recycled.

Polyester was discovered in 1941 by two chemists, Rex Whinfield and James Dickson, working for the Calico Printers' Association in war-torn Manchester. They heated together 1,2-ethanediol (aka ethylene glycol) and dimethyl terephthalate at 200 °C and got a sticky mass of polyethylene terephthalate which could be drawn into long, strong and stable fibres. These were particularly good in blends with cotton, producing a fabric which did not crease and was easy to launder. Polyester and polyester cotton now account for 80% of the clothing market.

In the 1960s a new use for PET was found in the form of transparent plastic bottles. They save energy, because they require only 45% of that needed to make a glass bottle. They save on transport, because a

[7] Some phthalate-free medical devices, such as those used for drip feeds, are now available and these use a new plasticiser called Hexamoll DINCH which is 1,2-cyclohexanedicarboxylic acid diisononyl ester.

delivery truck can carry 60% more soft drinks. And they do not break into sharp fragments like glass. While most such bottles are sold as non-returnable, it is possible to make them strong enough to be reused. Alternatively PET bottles can be recycled into different products by being melted down and turned into plastic that is suitable for other types of packaging, or into polyester fibre and turned into such products as carpets, duvets, anoraks, bristles for paint brushes and felt for tennis balls. Hulls for boats can be made from it, as can their sails.

Some PET is recycled chemically, in other words it is depolymerised into its starting materials by heating the plastic under pressure in methanol. The dimethyl terephthalate and ethylene glycol which this yields can then be purified, added to the feedstock, and returned for polymerisation to make virgin PET. Containers made from PET score over other plastics because of their ability to be air-tight, which is essential if colas are to keep their fizz. They are also used for sauces, cooking oil, honey, jam, cosmetics and paint. Medical products that need to be sterilised by irradiation are also best contained in PET, and it is also the preferred plastic for X-ray films.

Can PET be produced as bio-PET? This would require ethylene glycol and dimethyl terephthalate to be made from renewable resources. The former is made from ethylene so that is not a problem. The latter is made by oxidising *para*-xylene which would require a natural source of the chemicals from which this can be made, such as benzene.

Making benzene and xylenes from natural resources presents us with a problem, although clearly there are plants which can make these ring compounds, such as trees whose lignin component contains benzene rings with carbon atoms attached. A patent for producing benzene from this source was filed in 1981 by Hydrocarbon Research Inc of Lawrenceville, New Jersey. The process involved heating sawdust with hydrogen gas at around 400 °C and 100 atmospheres pressure, and the products included benzene.

Heating cellulose to high temperatures in the presence of a catalyst can produce benzene compounds as demonstrated by a group at the University of Massachusetts headed by George Huber who reported their findings in 2008. The cellulose first decomposes into a mixture of 300 liquid hydrocarbons and these are fed into the zeolite catalyst known as ZSM-5 at 600 °C. Within two minutes copious amounts of all kinds of chemical transformations occur including the production of benzene and the related xylenes.

Clearly PET has a future as bio-PET, but there is an alternative: bio-PEF. Instead of dimethyl terephthalate, it is possible to use 2,5-furan-dicarboxylic acid to make a polyester and this can be produced from 5-hydroxymethylfurfural (HMF), which forms when sugars and carbohydrates are thermally decomposed. Producing HMF can also be done in a one-step process as demonstrated by the bio-oil company KiOR based in Pasadena, Texas. The HMF can be oxidised to 2,5-furandicarboxylic acid and hence to PEF (aka polyethylene furan dicarboxylate) which is very similar to PET.

Furanics are based on the furan ring of four carbons and an oxygen. Furfural has an aldehyde (CHO) group attached to the ring and furanics were being produced from biomass such as corn cobs, oat husks and sugarcane bagasse early in the last century. This method of production finally ceased, but furanics could be revived and that's the hope of a company called Avantium which produces HMF. Avantium says it can produce biofuels and bioplastics from non-food biomass and at prices competitive with oil-based varieties.

There are phthalates and phthalates

You may now be wondering how a polymer made from one phthalate, *i.e.* polyester, can be so beneficial while another polymer which contains phthalate plasticiser can be so disapproved of. The explanation is simple: it all depends on the positions of the two acid groups which are attached to a benzene ring. If they are at opposite ends of the ring then the chemical is benzene-1,4-dicarboxylic acid (aka terephthalic acid) and this is the phthalate used to make polyester. If they are next to each other on the benzene ring then the chemical is benzene-1,2-dicarboxylicic acid and it is from this molecule that the phthalate plasticisers are made.

5.5 POLYSTYRENE (PS)

Best known as an insulator, but it has many other uses.

There are three types of polystyrene: general purpose, high impact, and expanded. General purpose polystyrene is used for drinking glasses, brushes, combs, razors, cosmetic compacts, disposable pipettes and DVD containers. High-impact polystyrene, which has been made stronger with rubbery polybutadiene, is formed into things like refrigerator door liners, coat-hangers, food trays and toys. Expanded polystyrene is best known as a lightweight foam ideal for thermal insulation, and is no doubt the most environmentally-friendly of polymers. Every tonne of expanded polystyrene insulation saves three tonnes of heating fuel a year. It also makes ideal packaging for electronic goods, crash helmet padding, and wet fish boxes. Unfortunately it floats on water and is blown by the wind, which is why samples of it are to be found disfiguring beaches around the world.

Expanded polystyrene allows actors to appear to be crushed beneath falling masonry. Even real-life concrete bridges and motorway flyovers are constructed with expanded polystyrene, which is added to create lighter structures but which retain their original strength. Expanded polystyrene is made by the inclusion of bubbles of gas, and this gives it its lightness and insulating capacity. The use of polystyrene as an insulator is discussed in more detail in Chapter 6.

Polystyrene has benzene rings bonded to every other carbon atom of the backbone chain. These make it hard because the benzene rings on one chain tend to attract those on another chain making the plastic tighter and stronger and this also results in a transparent material with a high refractive index, giving it the attractive sparkle of glass.

Polystyrene is recyclable, and that includes the expanded polystyrene which is used as protective packaging. It can be recycled into items such as plant pots, although a lot goes to the construction industry for concrete. The polystyrene which can't be recycled may end up in an incinerator but there it releases as much heat, weight-for-weight, as coal or oil. When polystyrene is accidentally ignited it can cause disastrous fires as it did in 1996 at Düsseldorf International Airport when 17 people died. A fire in the Channel Tunnel the same year again ignited polystyrene but thankfully no lives were lost, it took six months, however, to repair the damage and get the tunnel fully operational again.

Can polystyrene become a bioplastic? More than 24 million tonnes are made annually from ethyl benzene which is produced from benzene and ethylene. So in theory there can be bio-polystyrene, although none has yet been produced. Rather surprisingly, styrene occurs naturally in some vegetables, fruits and nuts but not at commercially exploitable levels, although this suggests that one day the gene that produces it might be transferred to another more productive plant.

5.6 POLYURETHANES (PU)

From ski suits to condoms, this is the sexiest plastic of all.

Polyurethane has proved itself to be a versatile polymer and today we find it in artificial sports tracks, ski suits, waterproof leisure wear – and condoms. Depending on the way it is manufactured, it is possible for polyurethane to vary from soft rubbers to hard solids. In the home, polyurethanes make life comfortable while in the car they improve safety. The main application for polyurethanes is foam cushioning for furniture and bedding, and for padded dashboards and steering wheels.

Polyurethane foam is flexible and is made by reacting isocyanates and alcohols with the addition of a little water. The water reacts with some of the isocyanate to create carbon dioxide (CO_2) gas, the bubbles of which get trapped in the viscous mixture of the two components as it polymerises, expands and then solidifies. Polyurethane foam is used to insulate the fuel tanks of space shuttles, although it was a metre-sized lump of this which broke off the external fuel tank of the *Columbia* and damaged the shuttle's thermal insulation, causing the craft to burn up as it entered the Earth's atmosphere on 1 February 2003.

Avanti condoms are made from polyurethane, and are twice as strong as the traditional latex ones, so that they can be made much thinner and therefore more sensitive, and they are transparent. They are non-allergenic, unaffected by lubricants and provide an effective barrier against sperm and all sexually transmitted diseases, including HIV. This kind of polyurethane, which incorporates an elastomer, is suitable not only for condoms, but wellington boots, the soles of training shoes, and for Lycra swimwear and hosiery.

Polyurethane can be used as an adhesive to bind together other materials, such as old tyres which are shredded and bonded to form athletic tracks and surfaces for children's playgrounds. And when its useful life is over, the polymer need not be wasted. It can be incinerated to release energy, or it can be separated and recycled by reducing it to its constituent chemicals which can then become a new batch of polyurethane.

Polyurethane is made by the reaction of a diol and a diisocyanate, and as there are several kinds of these molecules so there can be several kinds of polyurethane. Can there be bio-polyurethane? The diol component poses no problem. For example 1,3-propanediol can be produced from the fermentation of the carbohydrate dextrose. DuPont manufactures it from maize starch and uses it to make the plastic Sorona, although this accounts for only 30% of the polymer with currently unsustainable terephthalic acid accounting for the remainder. Sorona is used to make fibres for textiles, as well as packaging, and is very like nylon in its performance.

The diisocyanates pose more of a problem. The most commonly used ones have the two isocyanates attached to a benzene ring and they are made by first attaching amine (NH_2) groups to the ring, then oxidising these to nitro (NO_2) groups and finally to isocyanate (NCO) groups.

Again sustainability depends on a biomass source of benzene. There are polyurethanes made from non-benzene isocyanates but these have been less popular.

5.7 EXTREME POLYMERS

The polymers for which Nature has no use – but humans certainly have.

Polymers can be materials at the extremes: glass that never breaks; cord that never snaps; fibres that repel all dirt; body armour that stops bullets; clothing that repels water – and even packaging that dissolves in it. This last one is known as polyvinyl alcohol and it can be used to keep incompatible chemicals separate, that is until they come into contact with water, for example in a dishwasher tablet. In this section we will look at some of these remarkable plastics.

Polycarbonate is the unbreakable transparent polymer of CDs and DVDs, and it is used as a kind of safety window glass. The reason for its success lies with its incredible strength, durability, colourability, and the fact that it can be blended with other polymers. Yet there are those who doubt its safety because it is made from a chemical called **bisphenol A** which was shown to affect the growth of certain yeasts. However, research on rats and mice at high doses and over long periods of time showed no effects on the reproductive abilities of either sex. Indeed even when bisphenol A was given to pregnant rodents at levels that produced a toxic reaction, they still gave birth to normal babies. Nevertheless, such was the clamour against it that polycarbonate, which might contain minute traces of bisphenol A, was banned in some countries for uses such as baby bottles.

Although polycarbonate had been discovered in the 1890s, it was only in 1953 that the Bayer laboratories in Germany were able to produce material which was suitable for commercial applications, and it came on the market with names such as Xantar and Lexan. This was as clear as window glass and could be used as such. Polycarbonate remains stiff up to 140 °C and resilient down to –20 °C. It does not easily burn, and with the addition of flame-retardants it will pass severe flammability tests. It is not affected by most chemicals. Polycarbonate is naturally transparent, but it can easily be coloured and its surface embossed. Polycarbonate is used for Blu-ray discs, baby bottles, drinking glasses, light covers, safety glasses, visors, advertising signs, riot shields and instrument panels. The F-22 Raptor jet fighter has the whole cockpit canopy made of polycarbonate.

Can there even be bio-polycarbonate? The answer is yes, provided phenol, the fore-runner of bisphenol A, can be produced from biomass, and phenol is made from benzene. Acetone is the other component used in the manufacture of polycarbonate and this can be produced easily from all kinds of simple carbohydrate sources.

Acrylic is the general name given to another group of transparent polymers, of which the best known is the plastic **polymethyl methacrylate** (PMMA), better known as Perspex or Plexiglas. PMMA was discovered in Germany in 1928. At first it was expensive and only became widely available ten years later, after chemist John Crawford, working for ICI, found a method of making it cheaply from acetone. Perspex, as the new plastic was called, was ideal for cockpit canopies, illuminated signs, contact lenses, false teeth and artificial nails. It was transparent, almost unbreakable, and easily coloured, which is why it found favour with the furniture designers of the 1960s, especially for contoured chairs. PMMA transmits around 90% of visible light and it will filter out the more damaging UV light.

There are other acrylic polymers. Ethyl acrylate is also used to make them, and methyl cyanoacrylate is the basis of super-glues. The latter is sold as the monomer and has been designed by chemists only to polymerise when it is exposed to the air. Its superior stick-ability derives from the long chains of polymers that form across the surfaces it is joining, binding them together as if they were part of the same piece.

Over three million tons of acrylic polymers are manufactured worldwide each year, and they coat walls, floors, steel, paper and leather. It is also used as the binder for disposable non-woven fabrics, such as nappy linings and headrest covers in aircraft. The ethyl acrylate polymer is ideal for covering surfaces because it is flexible, tough, stands up well to regular cleaning, and has excellent resistance to bad weather and strong sunlight. It is used to coat the inside of aluminium cans so that their contents, especially fruit acids in drinks, do not react with the metal, and thereby contaminate the product.

Acrylic paints are used to cover metal surfaces, such as fridges, washing machines, dishwashers and cars, and they give them a lacquer-like finish. Indeed they are so strong that they can be applied to steel sheet before the metal is stamped into component parts. Acrylic polymers bind the pigments in paints and ensure they stick to the surface being decorated. Ethyl acrylate is added to a formulation if the paint is destined for surfaces where it must have a bit of flexibility, such as on textured wallpaper. In such paints the ethyl acrylate is co-polymerised with the more rigid methyl methacrylate which is the main component.

Generally the raw material for acrylic plastics is propylene, so bioacrylic plastics should be sustainable.

Teflon (aka polytetrafluoroethylene, PTFE) is the polymer to which nothing sticks. It was discovered by a 27-year-old chemist, Roy Plunkett at the DuPont research laboratories at Deepwater, New Jersey, in 1938. Its first major role was in the production of the uranium hexafluoride (UF_6). This required the handling of large volumes of fluorine gas which attacks virtually all other materials but not PTFE. The UF_6 was needed to make the atomic bomb which destroyed most of Hiroshima in August 1945.

Common Sense 13: It was thanks to going to the moon that we got non-stick frying pans

Wrong. The first non-stick frying pans went on sale ten years before Neil Armstrong set foot on the moon on 20 July 1969. When some people questioned the enormous cost of the moon flights, then $14 billion, the US National Aeronautics and Space Agency (NASA) pointed to the more down-to-earth benefits their research programme had brought, such as the non-stick frying pan. In reality, the moon landing would have been impossible without the plastic PTFE which was already being used to make non-stick kitchenware. The environments of extreme cold, low pressures and even the corroding effects of oxygen atoms in the Earth's upper atmosphere required a material stable to these conditions, and in the 1960s the obvious plastic was PTFE. Without it there could have been no voyage to the moon.

PTFE has some remarkable properties. It is not attacked by hot corrosive acids; it does not dissolve in solvents; it can be cooled to –240 °C without becoming stiff; and heated to 260 °C without affecting its performance. It has a low coefficient of friction which was to be the secret of its commercial success. So how is it possible to stick Teflon, to which nothing sticks, to the metal of kitchen ware? This technological triumph was achieved by Louis Hartmann back in the 1950s. He showed it was possible to bond PTFE to aluminium by first etching microscopic pits into the surface of the metal. The tetrafluoroethylene is then applied as an emulsion and when the metal object is heated the monomer polymerises into one continuous film of Teflon that is held tight to the surface by the polymer trapped in the millions of cavities on the surface. The French company that invented the non-stick frying-pan was Tefal.

Goretex is another PTFE product, best known for waterproof clothing and shoes. In 1969 Dr Bob Gore, of Maryland, found a way of expanding PTFE by heating and stretching the polymer to form a membrane. This creates invisible pores in the film – billions per square cm – and these are small enough to keep water droplets out, but big enough to allow water molecules of sweat to escape as vapour. Goretex film is sandwiched between the outer fabric and the inner lining and is widely used for wet weather gear and sportswear.

PTFE can make clothes, chair-covers and carpets stain-repellent. It does duty as plumber's tape for sealing joints in water pipes and central heating. At a more intimate level we use it in the form of dental floss. As you read this article your fingers will be picking up PTFE from the page. PTFE scrap from industry is re-used by grinding it to a microfine powder and then adding it to printer's ink to make it flow more smoothly. PTFE makes excellent fabric roofing for sports arenas and there is more about this in Chapter 7.

Has PTFE got a future? It has, as long as a fluoride-containing mineral, such as fluorspar (calcium fluoride), can be mined. Heated with sulfuric acid this yields hydrogen fluoride (HF) which is reacted with chloroform, a sustainable chemical which can be made by heating methane gas and chlorine. The product is **chlorodifluoromethane** ($CHClF_2$) and when this is heated to 600 °C, it forms the tetrafluoroethene from which PTFE is made. Fluorspar minerals are relatively plentiful and there will be little incentive to extract fluoride from sustainable resources, such as seawater in which the level is very low, around 1 ppm

Victrex (aka PEEK, short for **polyetheretherketone**) is a crystalline thermoplastic that can perform as well as a metal under certain conditions. It has incredible hardness, strength, wear-resistance, and is able to stand up to steam and high temperatures (250 °C), so much so that it can be used to make things like piston parts, bearings and pumps, and it is used in the chemical, aerospace and automotive industries. When glass fibres are added to Victrex it makes it more flexible and less sensitive to heat; when carbon fibres are added, it gets stiffer and stronger; and when carbon fibre and PTFE are added, then it has a very low coefficient of friction and is excellent for bearings. Victrex is also used in body implants and more than two million such devices have been inserted by surgeons.

Victrex could have a sustainable future. It is made from difluorobenzophenone and hydroquinone, both of which are made from benzene, so it has the potential to be sustainable. Indeed there are other ways of making hydroquinone, such as the one reported by a group at Michigan State University in 2001 which showed it was possible to synthesise it from glucose.

Figure 5.3 How does chemistry stop a bullet dead?

Kevlar is the plastic that can stop a bullet, which is why it is used in bullet-proof vests and protective face masks. On a weight-for-weight basis it is said to be five times stronger than steel. It protects planes by lining the engine compartment to limit the damage that might be caused should a turbine blade fly off, and because of its strength and lightness it is part of the framework of the Boeing 757.

Kevlar was discovered in 1965 by Stephanie Kwolek, working for DuPont, and it was launched in 1982. Fibres of Kevlar are used to make car and bicycle tyres much stronger and they are used to reinforce other plastics in boat hulls. Kevlar's strength also works against it, in that it is insoluble in all solvents, which means that working with Kevlar is rather difficult. However, it will dissolve in pure sulfuric acid, from which it can be extracted unharmed, and this is one way in which Kevlar can be processed. In addition to being almost immune to chemical attack, Kevlar is also fire resistant, flexible and lightweight. When it is spun into fibres and heat treated, the polymers get even stronger, and they are used for military armour, space suits, safety gloves and fishing rods. Kevlar is incorporated into tennis racquets, skis and running shoes, and these sporting aspects are discussed in Chapter 7.

Kevlar has one disadvantage and that is its need to be protected from direct sunlight, which causes it slowly to decompose and which limits its uses somewhat. However it makes an excellent outer coating for optical fibres, which are buried underground, protecting these from damage.

Few plastics have the package of benefits that Kevlar provides. When it fails, then it does so progressively rather than catastrophically, thereby providing another margin of safety. Unlike many plastics, it does not become brittle at low temperatures, even as low as –70 °C. Nor is Kevlar affected by long exposure to weathering or the sea, and three years immersed in either boiling water or in hydrocarbon solvent have shown it to be left unchanged.

Kevlar is flame resistant, self-extinguishing and gives off little smoke, so that it is the material of choice for conveyor belts, especially in mines, and for hoses used in the chemical industry and engines. Mooring ropes for tankers are made of Kevlar rather than steel, but perhaps the most dramatic use of Kevlar has been in body armour, flak jackets and head gear, which are not only lighter than other forms of protection but can be tailored to fit.

Could Kevlar be sustainable? The answer is yes, provided that benzene production can be made sustainable.

These then are some of the plastics on which modern society depends and they can have a sustainable future. There are many other plastics with special uses and they too are made from the same raw materials mentioned above. There are also plastics which were once manufactured but which failed to capture a market such as Carilon (aka aliphatic polyketone), which was made by polymerising ethylene with carbon monoxide – which means that it could easily be produced from biomass – and it had some rather remarkable properties. It combined the advantages of being easy to process, had superior resilience, high-impact performance over a range of temperatures and resisted chemical attack. It wore well and was ideal for plastic piping for corrosive chemicals, and for automotive fuel systems, including the fuel tanks, fuel lines and engine covers. Carilon was suitable for gear wheels in vacuum cleaners, cars and electric shavers, with the added advantage that they produced less noise and did not require lubrication. Carilon provided excellent barrier protection against hydrocarbons and oxygen, and retained its desirable features in an aqueous environment, hence it was seen as ideal for underwater off-shore cables and ties. Sadly Shell abandoned Carilon in 2000 when its corporate strategy changed and production ceased although it was aiming to produce more than 20 000 tonnes per year. My belief is that we have not heard the last of this wonderful and eminently sustainable material.

One day polymers might, like skin, heal themselves when damaged. Chemists have devised polymers in which tiny microcapsules of a monomer and of a catalyst are included in the polymer. If the polymer fractures at any point, these microcapsules break open and their contents flow to the site of the damage to polymerise and repair the injury. As yet no such self-healing polymers have been marketed but they might one day be made as a way of extending the useful life of plastics, and that of course conserves energy and raw materials.

That then is a brief resumé about our reliance on polymers and plastics and how we are slowly, but only very slowly, moving in the direction of sustainability. As we have seen, it is not an improbable goal to make all plastics sustainable when we stop using oil. The ultimate chemical plant will one day use only biomass and produce only biofuel and bioplastics, and require no external supply of energy and produce no waste. Maybe one country could show the way and an obvious one to do it is the UK. Could its population live a chemically sustainable lifestyle? This island of 60 million people has a land area of 24 million hectares. Of this 14% is urban, 12% is woodland and forest, 19% is under crops, 52% is grassland, and the remaining 3% is set-aside land.[8] Could it produce 60 million tonnes of food per year, 60 million tonnes of biofuels and 10 million tonnes of bioplastics? I believe it could. More importantly, it would set an example to the world that there is a sustainable future for humans. A hectare of land can yield around two tonnes of bioplastic which means that the UK would have to devote around 20% of its land area to produce all the 10 million tonnes of plastic that we require.

[8] UK Department of Environment and Rural Affairs.

CHAPTER 6

Cities and Chemistry

[A word in **bold** means there is more information in the Glossary.]

Cities offer employment, entertainment, education, economic wealth, excitement and close contact between people. In this chapter we will look at seven areas where chemistry is vital in making cities less wasteful and in making citizens feel better about themselves. The topics are: lighting, insulation, glass, solar panels, screens, laundry aids and personal hygiene.

Since 2008, all new homes in the UK have been rated according to their sustainability on a scale from 1 to 6 with these numbers referring to energy and water use. Level 1 requires the home to reduce energy demand by 10% and for its occupants to use only 120 litres of water per person per day. Level 3 demands a 25% cut in energy and limits water needs to 105 litres. Level 5 is 100% less energy and 80 litres of water, while level 6 is 'zero-carbon' although its water allowance is also 80 litres.[1] By 2016 all new houses built in England will have to be 'zero-carbon' and examples of such buildings can be seen at the Building Research Establishment in Watford.

Most supposedly green buildings are not really green. Their greenness is a token gesture of sustainability and often takes the form of using recycled materials. The energy saved is that which would have to be used to manufacture new materials, although the cost of recycled materials in energy terms may be more than people imagine. To make a house, office,

[1] *Code for Sustainable Homes, Technical Guide*, October 2008, Department for Communities and Local Government.

A Healthy, Wealthy, Sustainable World
By John Emsley

Published by the Royal Society of Chemistry, www.rsc.org

factory, or public building really green requires chemistry. The ideal green home is possible and its primary aim has to be energy efficiency; that's what the first five topics in this chapter are about. But living in a civilised society means meeting other people and feeling confident about oneself, and for that we need to take care with personal hygiene which means washing both oneself and one's clothes, and these depend very much on chemistry.

6.1 CITY OF LIGHT

Strange chemicals are making it possible to illuminate cities better.

Cities at night are ablaze with light. Indoors and out, we should aim to use as little electricity as possible for this purpose. In homes, the inefficient incandescent light bulbs are being phased out, and not before time, because these dissipate 90% of the electricity they consume as heat rather than light. Compact fluorescent lamps are now the preferred form of lighting. These last twenty times longer, emit the same amount of light, and use 80% less electricity. Eventually these will be phased out in favour of **light-emitting diodes** (LEDs) which use even less.

Figure 6.1 Cities will have a brighter and more efficient future thanks to chemistry.

When compact fluorescent lamps were first introduced around 30 years ago, they took several minutes to reach full brightness and they were very bulky. Today they are much smaller and some which employ higher discharge voltages reach full brightness almost immediately, which is necessary for lights in places like toilets. There are two components to a compact fluorescent lamp: the bulb itself, which is where chemistry comes into play, and the electronics which make it work, and for which we can thank other scientists.

In a fluorescent light there is a mixture of **noble gases**, mainly argon, along with a few milligrams of mercury. When an electric discharge passes through the gas it excites electrons of this metal and these emit UV light of wavelength of 254 nm. This activates the atoms of the coating on the inside surface of the lamp causing them to radiate visible light, which covers the range of wavelengths from 400 to 700 nm. This coating consists of phosphors which are compounds of **lanthanide metals**.

Compact fluorescent lamps are sometimes referred to as trichromatic[2] because they rely on phosphors which emit light in the blue, green and red parts of the spectrum, *i.e.* at around 450, 550 and 610 nm, respectively. Viewed together their light is perceived as white. The blue emission band comes from atoms of the metal europium, the green from a mixture of lanthanum, cerium and terbium, and the red from a mixture of yttrium and europium.

Street lighting also works on the same principle of exciting metal atoms, in this case of sodium which is why they emit the predominantly yellow light.[3] This can be made less starkly yellow by adding a little europium which provides a wider range of visible light. Sodium vapour lamps are also energy efficient and have long lifetimes, and modern versions of these work at higher pressures which allows other wavelengths to contribute to the overall visible range, producing a more pleasing light. Even so, they will be replaced by LEDs in the future, and two towns, Oldham and Rochdale, are where they will be first introduced. These advanced LEDs use 25% less electricity and have a lifespan of 50 000 hours.

Red LEDs based on aluminium gallium arsenide have been around since the late 1960s when they were commonly used in numerical displays. However, it was not until **gallium nitride** semiconductor was developed, with its large **bandgap**, that blue LEDs appeared and by combining red, green and blue LEDs it was possible to create what

[2] Trichromatic means three colour.

[3] Sodium has two powerful spectrum lines at 589 and 590 nm in the middle of the yellow region.

appeared to be white light. Alternatively, and more cheaply, is the use of blue LEDs in combination with a phosphor which converts it to white light.[4] Formerly gallium nitride was expensive because it had to be grown on sapphire wafers, but in 2009 a group at Cambridge University led by Colin Humphreys showed it was possible to grow it on silicon, reducing its cost by 90%.

The other benefit of LEDs is that they last a long time and some LEDs produced 30 years ago are still in service. LEDs are used particularly for small lights, such as decorative lighting, and they provide direct lighting as in flashlights and on vehicles. They make ideal destination displays for trains and buses. LEDs respond at full luminosity very quickly which makes them ideal for brake lights.

6.2 COSY CITY

Cities consume vast amounts of energy to keep people warm. Savings are possible thanks to chemistry.

Homes in the UK consume 27% of the country's energy. This could be much reduced by better insulation. In theory, if a room is perfectly insulated then the heat from its occupants should be enough to keep it warm, even in winter. However, there is no such thing as perfect insulation, but polyurethane and polystyrene come close. A truly environmentally friendly building has to be energy efficient and last a long time, and as such will require the products of the chemical industry to provide the insulation, ideally produced from sustainable resources. Insulation also needs fire retardants, coatings and adhesives but in this section we will focus only on the insulating polymers themselves.

Good insulation not only keeps homes warmer in winter, but air-conditioned rooms cooler in summer, and today's fridges and freezers work more efficiently all year round thanks to better insulation. We are intrigued when we hold a piece of expanded polystyrene to discover how warm it feels to the touch, and we appreciate that our home can be made more energy efficient with it. Loft insulation was the first way to improve home insulation and most houses have a deep layer of polystyrene for this purpose. Polystyrene beads are now added to concrete thereby greatly improving its thermal insulation without compromising its strength.

Neopor is the name given to two kinds of building insulation. The giant chemical company BASF makes its Neopor from polystyrene

[4] The phosphor is known as YAG, which stands for yttrium aluminium garnet, and it is doped with cerium.

granules which are converted to the large blocks of expanded polystyrene. This also contains flakes of graphite which make it even better as an insulator. A house whose outer walls contain this type of Neopor slabs will lose very little heat. The other kind of Neopor comes from a company of that name and it uses a rigid polyurethane insulation, in which there are minute cavities made from natural protein polymers extracted from the keratin of cattle horns, horses' hooves and crabs' claws. Neopor is incorporated into concrete in place of the usual gravel and it is known as cellular lightweight concrete which is not only stronger and lighter than ordinary concrete, but cheaper. A 10 cm layer of Neopor provides the same insulation as 50 cm of conventional concrete while at the same time using only 20% of the conventional materials which go into concrete, namely cement, sand and gravel. And this kind of Neopor is stronger than normal concrete thanks to the inclusion of polypropylene fibres in the cement mix. When this Neopor is struck with a blow strong enough to crack ordinary concrete it is merely dented because its minute bubbles of air take the strain.

BASF also makes a rigid polyurethane insulator and their brand is called Styrodur. Its tiny cavities are formed using air. It is ideal for insulating buildings and for things like biogas generators. These work best at slightly higher temperatures, the heat coming from the fermenting slurry itself, and if they are properly insulated then no external heating is needed. Styrodur is impervious to water so that it can be used to insulate these generators which are sited partly underground.

The above insulation can be incorporated into new homes and apartment buildings but what about existing homes? There the obvious answer is cavity wall insulation. Twelve million UK homes were built in the last century with an outer and an inner wall, made of brick and breeze block, respectively, with a gap between. This was to prevent dampness penetrating from the outside. Today about half of them have this gap plugged with better insulation, which generally consists of a mixture of mineral wool and expanded polystyrene and this is formed within the cavity itself. Cavity wall insulation pays for itself within four years through reduced energy bills. Nor does it breach the damp-prevention of the cavity because the polymer is waterproof.

Chemists have another material with which to insulate rooms. It can be used on internal walls so that they act as storage heaters, absorbing heat when rooms are warm and releasing it when the room cools down. BASF's Micronal or DuPont's Energain are this kind of product and they are based on the method used to keep astronauts' spacesuits and soldiers' battlefield outfits at an even temperature. These incorporate paraffin wax which cools the wearer down by absorbing heat when things get too hot for comfort, and releasing it when the occupant starts to feel cold.

Micronal has micron-sized capsules made of polyacrylate, inside of which is paraffin wax. These are too small to be damaged even by such rough treatment as sanding and they are supplied either as a dry powder or dispersed in water. They are mixed with plaster to make plasterboard or are applied as wet plaster to a wall or as screed to a floor. They can even be added to paint. A layer of plaster half an inch thick (1.25 cm) containing 30% of Micronal is equivalent to a six inch brick (15 cm) in terms of storing energy. The Micronal microcapsules begin to work when the temperature reaches 26 °C and they absorb energy as the wax melts. The energy which 1 kg of these capsules will absorb in melting is equivalent to that needed to raise a litre of water from room temperature to boiling point.

Energain also uses paraffin wax. This is trapped inside a polymer matrix which is sandwiched between aluminium sheeting. The wax is designed to melt at 22 °C and as it does, it too absorbs a lot of heat while remaining at an even temperature. When the room begins to cool, the wax cools and at 18 °C it begins to solidify again and now it releases the same amount of heat as it absorbed. Energain panels are about the same size as conventional plaster board panels, *i.e.* around 1 metre square, and they are half the weight of normal plaster boards. Incorporating them into the walls of a room can result in energy savings of 30% or more. A single panel can store around 1000 kJ of energy, so that a room whose walls and ceilings contain 40 such panels, will be able to store the equivalent heat of a 1 kW electric heater working for 10 hours.

6.3 CITY OF GLASS

Glass is need for admitting light but it can be a liability when it comes to heat loss and cleaning. Chemistry has the answers.

Cities are full of windows and the glass of which they are made has many faults. Glass technologists and chemists have produced some remarkable materials in their time, such as glass which will even stop a bullet, but what we really need now is something which is more difficult to stop, and that is the flow of heat. We want windows to stop heat escaping from the building in winter and prevent heat entering in summer. Even better would be for windows to convert sunlight into electricity, and also to clean themselves.

Sageglass was the invention of John Van Dine and it has a coating which darkens when a direct current voltage is applied across it and the

Figure 6.2 Modern cityscapes gleam with glass that plays an active role in making buildings more energy efficient.

voltage can be as little as 5 volts. It is ideal for skylights and other windows which might bear the full brunt of the sun's rays. A flick of a switch and Sageglass windows darken by forming a mirror of lithium metal. The reflective layer is applied to the inside of the glass and consists of two transparent conducting electrodes, one applied to the glass and the other providing the outer layer. It is across these that the voltage is applied. Sandwiched between them is a layer containing lithium ions (Li^+), and they can pick up electrons from the electric current and become silvery lithium metal atoms. These reflect 90% the heat of the sun while still allowing some visible light to pass through. Sageglass layers are applied to float glass and they are only 5 microns thick (about 1/50th the thickness of a human hair).

PowerGlaz is the name of the new architectural material designed for use as facades and roofs of buildings. PowerGlaz panels contain an array of silicon-based solar cells laminated between two layers of glass, and they can deliver around 900 watts of power when exposed to the midday sun.[5] The panels are manufactured at the Romag factory at Consett in County Durham, which has the capacity to make 7000 such panels a year, producing a total of 6 MW, and they guarantee a power

[5]The solar cells are 12.5 cm (6 inch) square and the panels are 3.3×2.2 m (11×7 ft) in size.

output of 80% over a 25 year period. They can be used for solar farms, as well as being mounted on buildings. The International Business Centre at nearby Gateshead and City Hall in London are just two of a growing number of public buildings now fitted with PowerGlaz panels. The International Business Centre has 36 panels capable of generating enough electricity to meet the building's needs on a sunny day.

PowerGlaz also makes PowerPark, which is a canopy of PowerGlaz panels that will be used for the parking areas of places like airports, supermarkets, stations, offices and anywhere where large numbers of cars are parked. Such a solar parking facility already exists at the California State Fair where 1000 parking lot canopies are covered with solar panels, which produce 0.5 megawatts (MW) from sunlight that would otherwise be roasting vehicles. There is a certain irony in generating solar energy by providing shade for parked cars, which are probably the world's biggest wasters of energy. PowerPark will be particularly beneficial when electric vehicles become popular so that their owners can recharge the cars' batteries from power generated on site.

Glass can control the amount of heat it lets in to the building during the day and the amount that is lost from the building at night, if it is coated with a 0.3 micron film of tin dioxide (SnO_2), plus a little fluoride. Such a layer lets visible light through but reflects infrared heat back and this kind of coating is applied to the glass surface as part of the manufacturing process.

The trouble with glass is that it has to be cleaned, especially in urban environments where city dust and vehicle emissions can settle on it. The chore of cleaning windows becomes a less frequent activity when these are clad in self-cleaning Activ glass, produced by Pilkington and launched in 2002. This is glass whose molten surface has been exposed to titanium tetrachloride ($TiCl_4$) and steam, and these react chemically to form a layer of titanium dioxide (TiO_2) only 50 nanometres thick and which is permanently bonded to it.

So how does it work? The answer is that the TiO_2 absorbs energy from the sun to activate an electron which then attaches itself to a molecule of oxygen from the air to create the superoxide free radical $O_2^{\bullet}$. This will oxidise any organic dirt like grease on the surface of the glass, converting it to CO_2 gas. Any inorganic dirt left behind is then washed away more easily because the surface layer of titanium dioxide retains a film of water, unlike ordinary glass on which rain gathers as droplets. The result is that when rain hits Activ glass it washes its surface clean again, and is ideal for tall buildings with glass façades.

An alternative kind of self-cleaning window was launched by Henkel in 2004. This window makes use of silica nanoparticles which form an invisible film with a negative charge. This makes the glass attractive to water so that it retains a water film on its surface on which the dirt deposits, only to be easily washed away by rain.

6.4 THE CITY AND THE SUN

Cities consume the most electricity and with a little a help from chemistry they might well produce more for themselves.

On a global scale, the electricity generated in thousands of power stations is enormous. Even so, it pales in comparison to the amount of energy the Earth receives from the Sun. There is enough sunlight falling on the Earth in one hour to supply all the world's energy needs for a year. We can harvest some of this energy by using solar panels, but at the present time these generate only a tiny fraction of the 20 000 billion kilowatt hours (kWh) which the world needs. Our reluctance to plug into this 'free' energy is paradoxically due to its high cost: the energy generated by solar panels is several times more expensive than energy generated in other ways. Despite this, the world is investing in solar energy and the current target is to reach at least 15 billion kWh by 2010. (The US Government's Energy Information Administration predicts that global demand for electricity will rise to 30 000 billion kWh by 2030.) Energy from solar panels is unlikely to account for more than a few percent of this total.

Generating electricity and getting it to where it is needed currently consumes vast amounts of fossil fuel, and a third has leaked to the environment by the time it reaches its destination.[6] Ideally we should aim to generate electricity from sustainable sources of energy and as near to the point of use as possible. Solar panels hit both targets, but Nature isn't going to make it easy. For a solar cell to convert sunlight to electricity requires a semiconductor, which can absorb the sunlight to generate negative electrons and positive 'holes'. These need to move in opposite directions thereby producing an electric current, and this is what happens at a semiconductor diode junction. The result is **photovoltaic (PV) power**.

The PV effect was first demonstrated by a 19-year-old Frenchman Alexandre-Edmond Becquerel in 1839, but it remained little more than a scientific curiosity for the next hundred years or so. The first solar cells

[6] Chemists might one day find a superconducting material which will conduct electricity with no loss of power and which would work at normal temperatures. Currently superconductors have to be cooled to very low temperatures, such as –250 °C, to operate effectively.

appeared in the 1950s and they were fitted to satellites put into orbit round the Earth. They were made of pure polycrystalline silicon and they could convert 4% of the sun's rays to electricity. There are two kinds of efficiency of PVs: quantum efficiency and power efficiency. The former is the efficiency with which a material converts photons of light to negative electrons and positive holes, while the latter is the real efficiency that can be extracted as electric power and this is far less.

Silicon still accounts for around 90% of the solar cells in use around the world, and they now have conversion efficiencies of about 15%. Japan produces 95% of suitable grade silicon and the reason the industry flourishes there is that the Japanese Government decided to invest in solar energy following the Oil Crisis of the early 1970s, when its price increased dramatically.

First-generation PVs, of which there are many, are made of polycrystalline silicon, but their efficiency in turning sunlight to electricity is relatively low. Second-generation, thin-film cells are made from alternative materials, such as copper indium diselenide (aka CIS) and especially cadmium telluride.[7] Third-generation cells use a combination of semiconductors, as in the multi-junction tandem solar cells. These typically have a top layer of gallium indium phosphide, a middle layer of gallium arsenide and a lower layer of germanium, and they are able to harvest sunlight across more of the sun's spectrum and achieve power efficiencies of 40%. In June 2005 the company Spectrolab, a subsidiary of Boeing based in Sylmar, California, was claiming this level of efficiency for its multi-junction cells. Third-generation cells can also maximise their output by concentrating the rays of the sun and tracking its path across the sky.

Japan aims to meet 50% of its electricity supply from solar panels by 2030, when there should be an array of them on most roofs. They might well achieve this because they have some of the largest solar cell manufacturers in the world. Sharp leads the field, and there are other manufacturers like Sanyo and Mitsubishi. German companies Q-Cells and Schott Solar are also big producers. Meanwhile the UK generates only a minute amount of its electricity this way, but we are situated on a northern part of the globe and the sun's rays are often hidden by clouds. As we shall see, these disadvantages can be overcome.

Like all forms of power generation, solar energy systems are prone to being somewhat less than totally reliable, as events in Spain have shown. In 1994 an array of 5000 solar cells was set up at the village of La Puebla

---

[7] Cadmium products are not permitted to be used in Japan on grounds of toxicity.

de Montalbán in the central region, designed to feed the national grid with 1 MW of power. In fact it delivered 0.85 MW. During the 10 year trial only 50 solar cells broke down – and a few were stolen. Some loss of power was inevitable in any case, due to the fact that semiconductor grade silicon decays by about 1% a year. Solar cells need to be connected in series to generate power for practical power applications, and the current they produce is direct current but this can easily be turned into alternating current.[8]

Silicon accounts for about 60% of the cost of current solar cells.[9] Given that silicon is the second most abundant element in the Earth's crust – oxygen is the first – why is it so expensive? The PV industry used to use the silicon which was not wanted by the microelectronics industry. Recently the growth in the PV industry has meant that its demand for silicon exceeds that of the microelectronics industry so it has had to develop its own silicon. What it now uses is so-called metallurgical silicon, which is cheap but then has to be purified in order to make the silicon needed for PV cells. Tiny amounts of phosphorus or boron also have to be incorporated into the silicon before the whole is cast into large ingots weighing 240 kg. These ingots are cut into 25 brick-shaped blocks and each is sliced into 400 wafers (250 microns thick) using a fine cutting wire so that each ingot eventually produces 10 000 of them.

Research by chemists and material scientists is focussed on alternative materials for photocells and is based either on elements which are more efficient than silicon but dearer, or on organic semiconductors and carbon nano-structures, which are much less efficient but far cheaper. Each approach has its attractions and its drawbacks. Metal-based semiconductors are potentially much better PV materials than silicon, because they absorb sunlight more efficiently and can also more closely match the energy of the incoming light rays. Gallium arsenide, indium phosphide, and cadmium telluride are the best ones, and the last of these has made significant inroads into solar power generation. The material which might overtake even this, in terms of widespread application, is copper indium diselenide (CIS).[10] A cell made of CIS at the US National Renewable Energy Laboratory in Golden, Colorado, has achieved an energy efficiency of almost 20%. Replacing about a quarter of the indium in CIS with gallium, not only reduces the amount of the more expensive metal, but improves the efficiency. The US company Nanosolar began commercial production of these cells in 2007.

[8] Direct current is more efficient but society is geared up to alternating current at the present time.
[9] Other materials account for around 20%, interest payments for around 15% and labour for 5%.
[10] Chemical composition: $CuInSe_2$.

Thin-film PVs promise to be a major contributor to electricity production. Those based on silicon use amorphous silicon which is radically different from crystalline silicon in that it does not need to be cast into ingots and it can be applied to a substrate, such as glass or, better still, a flexible material so that the completed panels can be moulded to any surface. Thin-film photovoltaics are now rolling off the machines at various factories around the world.

The alternatives to silicon and metal-based semiconductors are organic semiconductors of the kind mentioned earlier, but they currently struggle to reach efficiencies of 6%, although there are hints in the scientific literature of materials that will enable much higher PV cell efficiencies.

When the sun shines all is well for solar cells, but what happens when the sky is overcast? The answer might be the dye-sensitised cell which can operate using only daylight. Michael Grätzel at the Swiss Federal Technology Institute of Lausanne made the first such cell in 1991; it relies on a compound of the rare element ruthenium capturing a photon of light and using this to kick electrons into action. A dye-sensitised cell is efficient as a solar cell because it can absorb light coming from any direction, and it is particularly good at absorbing the blue end of the spectrum, which is the light that best penetrates clouds. Ruthenium is expensive but it can be replaced by copper according to research at Basel University in Switzerland, although their cells have only achieved an efficiency of 2.3%, whereas ruthenium ones manage almost 10%.

In dye-sensitised cells the light is captured by the metal-containing molecule, which then releases electrons into a metal oxide and 'holes' into an electrolyte and, provided they successfully migrate to opposite electrodes, they will produce an electric current; if they recombine before doing so they produce no current. To be effective, charge collection at the electrodes has to be much faster than recombination within the electrolyte. Arthur Frank of the National Renewable Energy Laboratory in Colorado found a way of improving dye-sensitised solar cells using TiO_2 nanotubes into which the dye molecules were placed. Wasteful recombination was 10 times slower, so the cells generated significantly more current.

Concentrated power PVs can have an efficiency of more than 30%. Such systems use lenses to focus sunlight onto a much smaller amount of photovoltaic material, and the lenses used are Fresnel lenses which are flat with miniature circular grooves which focus light on to a central area. In a well-designed cell the light falling on an area of 100 cm^2 can be focussed on to 1 cm^2 of photovoltaic material. The main attraction is that they need less of the solar cell material, which means that these can

Common Sense 14: If every roof had solar panels then our energy problems are solved

Maybe. Solar panels will probably never generate all the energy a modern city needs and in any case the world does not have enough indium metal to make this possible, if current solar cells are the only ones available. Indium is needed for **indium tin oxide** which has three essential features: it bonds strongly to glass; it is transparent; and it conducts electricity. It is widely used in solar panels and other kinds of electronic equipment. The problem is that there are no indium mines on this planet nor ever likely to be. What indium we have comes from mining zinc and lead, whose ores contain a little of this metal element. Chemists will have to discover an alternative to indium tin oxide and some have been found, but so far they have not been widely used.

Unsustainable metals?

Organic chemicals derived from carbon are sustainable because biomass is constantly being produced. However, when it comes to certain elements then the supplies are limited. According to Harald Sverdrup of the Department of Chemical Engineering of the University of Lund, Sweden, this century we are likely to run out of helium, silver, gold, tin and zinc, assuming we continue to use them the way we now do, while platinum, lithium and gallium will run out in the next century. Some of these are easily recycled because they are regarded as precious and no doubt will always be regarded so. Even aluminium might eventually become scarce as the workable deposits are used up, and while supplies of uranium ore are also depleting there are still large deposits of thorium which can not only be used as nuclear fuel but generated more fuel as it is used. Supplies of thorium are enough to keep such nuclear power stations in business for 25 000 years.

be made from expensive semiconductors like gallium arsenide. A light concentrator has been developed by Sharp and Diado Steel of Japan that focuses cells which are 7×7 mm in size, and achieves a conversion of more than 35%. Whitfield Solar is a company set up by George Whitfield of the University of Reading, England, and his concentrated power PVs are silicon-based and designed specifically with low cost and ease of manufacture in mind.

Much is promised from solar PVs and there are already many mega arrays around the world. For example, Portugal has its remarkable Girassol ('sunflower') PV power station at Moura which has 350 000 solar panels, covering an area of 112 hectares (equivalent to 168 football pitches), and generates 62 MW of power capable of serving the needs of 20 000 homes. One day there may be enough PV panels covering the roofs and sides of buildings in cities to generate 10% or more of the electricity which the city requires. It is possible.

6.5 INFORMING CITIZENS: PICTURES AND MOVING IMAGES

Citizens spend much of their day looking at images on screens. What has that to do with chemistry?

The camera with film is now the province of a dwindling band of dedicated traditional photographers. The rest of us prefer digital image making which has the advantages of immediacy, plus the ability to manipulate the images as we wish. Photography relied very much on chemistry, whereas the newer methods rely on physics and information technology. However, chemistry still has a role to play in the form of semiconductors, phosphors, liquid crystals and plasma cells.

A digital camera focuses light on to a semiconductor device which records it electronically. Image quality depends on the size of the minute components which make up the picture and ideally these must be too small for the human eye to distinguish so that the image appears continuous. These components are referred to as pixels,[11] and the more there are, the greater the resolution. A top-of-the-range digital camera may have as many as 10 million pixels (10 **megapixels**) although this is still an order of magnitude less than the pixels of the human eye which has 100 million pixels.

A pixel in a digital camera is a diode which can either respond to light or not respond. In itself, this would produce only a black and white picture, albeit one in which there were many shades of grey. But we want more than this, we want coloured images. So a beam-splitter directs the light to three different sensors and each is designed to respond to only one specific primary colour. Alternatively, and more economically, a colour filter array is placed over each photosite and this generally takes the form of rows of red, green and blue filters and there are as many

[11] Pixel is short for pix element, pix being the common abbreviation for picture.

Figure 6.3 Tap the right key and molecules will dance for your enlightenment.

green filters, as blue and red combined. This is done so that what the eye finally perceives is the true colour. The computer in the camera then converts the mosaic of information into the final picture.

Many of the images we encounter today are moving images on flat screens used for home viewing and public displays. Flat screens rely on chemistry in the form of liquid crystals, noble gas plasmas, LEDs and OLEDs. A liquid crystal display is an array of pixels which create moving images using very little power, and it is for this reason that they are in laptop computers and other devices which rely on batteries, and they make slim screens possible. Whenever we view our laptop we are seeing liquid crystal molecules twisting and turning as they respond to electrical impulses.

The term **liquid crystal** sounds to be an oxymoron, like living dead. This analogy can be carried a little further; what liquid crystal refers to is the intermediate state between the stationary (dead) arrangement of molecules in the solid phase and the mobile (living) condition of the liquid state. The transition between these two states occurs at a specific temperature known as the melting point.[12] In 1888 an Austrian botanist Frederich Reinitzer found that the solid phase of cholesteryl benzoate

[12] The melting point was once commonly used as a way of identifying a substance.

had *two* melting points. When its crystals were heated they melted at 145.5 °C to form a curious cloudy, viscous liquid which on further heating suddenly changed again at 178.5 °C to give the clear fluid of the liquid state. He told a physicist colleague Otto Lehmann what he has seen and Lehmann then observed the changes under a polarising microscope. He saw that something strange was happening: the cloudy viscous liquid exhibited a range of colours. It was Lehmann who described them as *fleissende Kristalle* (liquid crystals). At the time they were merely a chemical curiosity with no apparent application; today they support a billion dollar industry.

For a substance to exhibit liquid crystal behaviour it has to consist of molecules in which there is a rigid head attached to a flexible tail. The man who really developed liquid crystal technology was chemist George William Gray, at Hull University. He made new kinds of liquid crystals and especially ones which could perform their magic at room temperature. His liquid crystals consisted of a short hydrocarbon chain attached at one end to two co-joined benzene groups, and at the other end to a cyanide group. This was to open the route to liquid crystal displays (LCDs) and this compound, along with similar liquid crystals, is now used in nearly all such devices. When a voltage is applied across the liquid crystal screen then molecules line up and thereby allow a beam of light to pass through them.

So what is required of a liquid crystal with commercial applications? It should have a wide temperature range, ideally from −30 °C to +50 °C, so that the device will work in the cold as well as at normal temperatures. It must be stable chemically and remains so while it is experiencing an electric field. Its molecules must respond very quickly to changes in voltage so that moving images can be viewed. In fact, no simple liquid crystal meets all these requirements, which is why LCD devices contain a mixture of as many as 20 molecular variants.

Liquid crystal displays meet the ordinary citizen's needs but they have their limitations, namely they are best viewed full on, meaning they are not really suitable for group viewing. For that we need plasma or OLED screens and they are based on the chemistry of millions of phosphorescent cells sandwiched between plates of glass, with each cell containing a mixture of neon, argon and xenon gas. The benefit of plasma screens is that they can be viewed from all angles. When a cell is excited electrically, it creates a plasma whose intense glow then excites phosphors that emit red, green and blue light. This is done *via* rows of electrodes along the front and rear glass plates, the rows being positioned horizontally on one plate and vertically on the other so that when an electric current flows it will only activate those pixels where the electrodes cross. The screen to be viewed has an indium tin oxide as the conducting layer. A pixel will receive a voltage pulse several thousand times a second.

A plasma is created in a gas when a high electrical voltage is experienced, which pumps electrons into the gas. When these electrons collide with the atoms of the gas they excite those atoms, kicking their electrons into higher energy orbits. It is when an electron returns to its normal orbit that light is emitted. The gas which gives the brightest light is xenon. More importantly the plasma also releases UV light which can be used to excite phosphors, thereby generating more colours. The gas is contained in hundreds of thousands of tiny pixels sandwiched between two glass plates and there are plasma screens which are now very thin. Bang and Olufsen's BeoVision 9 has a 50 inch plasma screen for home cinema.

There is even the possibility that one day **organic light-emitting diodes** (OLEDs) will be widely used as flat screens. These do not need special metals, so in theory they are sustainable. OLEDs are based on organic semiconductors. The Nobel Prize for chemistry in 2000 was awarded to the discoverers of conducting polymers: Alan Heeger, Hideki Shirakawa and Alan MacDiarmid. Diodes made from organic semiconductors give different colours and there is a complete palette of hues to choose from. MP3 players use OLEDs because not only do they give vibrant colours, but they draw much less power from the battery compared to liquid crystal displays, which have to be illuminated from the back. But can OLEDs last as long as conventional LEDs? To date the longest-lived OLED panel has lasted 10 000 hours, but the goal is to devise one that will last 100 000 hours which would be 11 years of continuous output.

OLEDs are the most efficient way of converting electricity to light and this is why they are used in battery operated devices, and they are cheap to manufacture. OLEDs can generate light of whatever colour is desired, and the organic material can be applied by printing techniques rather like those of the inkjet printer. Currently the practical application of OLEDs are as display screens, some MP3 players, mobile phones and the Sony TV known as XEL-1. As OLED prices come down in the next few years they will become more common, and what may give OLEDs a great future is that they can be printed on to all kinds of surfaces, even fabrics.

Those then are the various ways in which chemistry can contribute to life in the modern city insofar as it concerns energy. Citizens also require a great deal of energy in the form of hot water with which to clean themselves, their homes, their dishes and their clothes. In hot countries, hot water can be generated using solar heating devices situated on rooftops. For many in the developed world, the issue is really one of how to reduce the amount of hot water that we need. In Chapter 2 we saw that using a dishwasher is beneficial in this respect and when it comes to washing ourselves, then we have the choice of taking either a shower or a bath. The former uses less water than the latter, unless it is a power shower, and consequently consumes less energy. When it comes to washing clothes then chemistry can really help reduce the demand for hot water.

6.6 CLEAN CLOTHES

'People watching' is a part of city life, so we want to look good and that means clean clothes.

If humans had evolved to be fur-covered animals, like other primates, there would be no need for clothes or for the 22 million tonnes of detergents used annually to wash them. People today wash clothes much more frequently than previous generations and while this may still be done to remove dirt and stains, sometimes it is done simply to refresh them.

With a modern detergent it is possible to wash clothes, towels and bedding, using very little energy, and relatively little water if a front-loading machine is used. For most people, doing the washing is now as effortless as putting things in the machine, priming the dispenser with detergent and fabric softener, choosing a wash cycle, and an hour or so later the job is done, or in less than half this time if a quick wash cycle is chosen. Clothes are now washed so frequently that in many families the washing machine is in use every day.[13]

Manufacturers offer a range of detergents, some of which have 20 or more ingredients. The major ones are surfactants which lift off dirt and grease, enzymes which digest stains (or bleaches to remove them chemically), water softeners which stop limescale forming, and chemicals like **carboxymethylcellulose** which prevent dirt from reattaching itself to fabrics.

Laundry detergents improved greatly during the last century and they are now much more environmentally friendly. There are three approaches to better detergents and different manufacturers have chosen different routes. One is to use only chemicals from sustainable

[13] A survey conducted by Unilever in 1997 found that the average family does about 5 laundry washes per week.

resources, but as yet such products do not perform as well as the traditional ones. An alternative route is to continue with traditional ones, but to provide them in a concentrated form which saves on packaging and transport. The third way is to provide detergents which will wash clothes even in cold water. Thirty years ago most people washed their clothes in very hot water at 80 °C and washing machines consumed a lot of energy. Ideally all three approaches will come together some time this century, meanwhile the best one appears to be washing in cold water. An example of this kind is Proctor & Gamble's Ariel Excel Gel which works superbly well in water as cool as 15 °C. Its surfactants come in part from sustainable resources. If all washes in Europe were done in cold water then it would save the output of 10 power stations, and in the US it would save energy to the value of around $6 billion per year.

If a load of washing is basically white or light coloured, then a detergent containing a bleaching agent, such as sodium perborate or percarbonate, may be chosen. While this works best in hot water it needs a bleach activator if it is to work at much lower temperatures, such as **TAED**. The stains on clothes and table linen which come from red wine, tea, colas, beetroot and fruit juices are easy to remove this way.[14] Professional laundering, such as that for hospitals and hotels, can use ozone as the bleaching and sterilising agent. The UK company JLA markets its OTEX device which destroys all microbes and which generates it ozone from the oxygen of the air. More than 2000 have already been installed.

If the wash load is highly coloured, then bleaching agents are to be avoided because these may attack dye molecules and colours will fade. The best way to remove stains in these instances is with enzymes. There are various kinds and these remove carbohydrate, fat and protein residues, including the unmentionable ones on underclothes. They break such residues down into smaller, soluble fragments. Enzymes are polypeptides and special varieties have been developed which will even work well in cold water. Some enzymes are now even compatible with bleaching agents, whereas formerly they were regarded as incompatible. Today, enzyme demand is in excess of 30 000 tonnes in Europe alone and most of this ends up in detergents. Not that this has prevented a rather curious delusion about enzymes persisting in the UK where many people deliberately avoid them.

---

[14] The most difficult stains to remove are curry, coffee, mud, motor oil, grass and grease. Only the best detergents can remove these so-called 'torture test' stains.

Common Sense 15: It's best to buy non-biological detergents, which are enzyme-free, because the enzymes in detergents cause eczema

Wrong. It is only in the UK that 'non-bio' detergents are sold thanks to a misleading BBC television programme. This gave air-time to a group of people who complained in the early 1980s that Persil New System, an enzyme-containing detergent, was causing them to break out in rashes and other skin complaints such as eczema. They blamed the enzymes but there was no scientific evidence to support their claims – and there never has been. Yet this belief has persisted and non-bio detergents still have around a 25% share of the market. Nowhere else in the world are such products sold, but maybe the answer is that British skin is more sensitive!

There are less obvious ingredients in detergents. There is often an optical brightener which absorbs UV light and emits it as white lights so that white laundry look whiter, thereby countering the yellowing of cotton and other fabrics which comes with age and frequent washing. There are foam regulators, to prevent a build-up of foam which can overflow from the machine. Fragrances are also added to detergents, designed to impart a pleasing smell to washed clothes, towels and bedding. And if you are worried that some of these chemicals might affect you then there are fragrance-free detergents. There are dye transfer inhibitors, to prevent dye molecules from colouring other clothes.[15]

Whenever something new that is highly coloured is washed, such as a T-shirt, a lot of the dye may end up in the water. If that garment is washed along with white or light-coloured clothes then there is a real risk that they will pick up some of the dye and emerge slightly coloured and be ruined. One answer to washing dyed fabrics is to separate the laundry into lights and darks, and wash them separately. Even so this does not entirely solve the problem of dye transfer and it was only in the 1990s that chemists came up with an answer: PVP, short for poly (*N*-vinylpyrrolidone). This polymer is water soluble and along its chain there are chemical groups that attract and hold on to those dye molecules that have floated off into the wash water, thereby preventing them from attaching themselves to other fabrics.

How do detergents fare in respect of the Renewable Carbon Index (RCI)? This is the ratio of renewable carbons in a product compared to the total carbon content of the product. Ideally this should be 1, for

[15] There are also various pre-wash products to treat difficult stains.

some eco-products it is 0.9, while for the best-selling detergent brands it is still below 0.25, although it is creeping up. Ideally the carbon content of their ingredients should all come from renewable resources and no doubt this century they will. The chemistry to achieve this goal is already understood and it is only the cheapness of oil-based resources which prevents it happening at the present time.

Finally, a word about a new way to wash bed linen and towels for hospitals and hotels. A washing machine has been devised which uses nylon beads to remove dirt and stains. The method has been devised by Stephen Burkinshaw, Professor of Textile Chemistry at Leeds University, and it uses thousands of nylon beads and only a few litres of hot water. The company Xeros has been set up to produce the new machines and they have two drums, one in which the wetted laundry is scrubbed with the beads, the other in which the beads are reclaimed for further use. After a hundred or so cycles, the nylon beads have absorbed all the dirt they can and these are then re-cleaned by using a vacuum process. Energy saving with the new washing machines is around 30% and water saving is in excess of 70%.

6.7 CLEAN CITIZENS

Personal hygiene fails when chemistry is ignored and in cities it can spell social disaster.

Finally we come to the more personal topic of body odour. Living in a city generally means coming into close contact with other people, sometimes in hot and sweaty environments, such as crowded trains, bars and night clubs. Chemistry's role in those situations is not so much to do with saving energy, more to do with saving face.

If cleanliness if next to godliness, then citizens today should have no difficulty in being in a state of grace, provided they have taken advantage of the products which chemistry provides to make this possible. No one need now fear that they are emitting off-putting body odours, but it was not always so. "The great unwashed" was how the Victorian novelist Edward Bulwer-Lytton referred to a distinguishing feature of the lower classes. Today everyone can be clean and fresh thanks to detergents, shower gels, antiperspirants and deodorants.

Our skin is protected by an oily layer called sebum to which stick dead skin cells, dirt and microbes. Most of us wash these away with a shower gel. This contains surfactants produced by the chemical industry, ideally from natural and sustainable resources. The first commercially successful surfactant was sodium alkyl benzene sulfonate but it was too

Figure 6.4 A little chemistry can refresh you and prepare you for the day ahead.

powerful for personal use. Later surfactants were devised which were much gentler, such as the sodium laureth sulfate you see as an ingredient in many products. Even newer surfactants are the alkyl glycosides which are made from renewable resources such as carbohydrates and plant oils. These are not only gentler to the skin but they foam well with a creamy feel to the lather.

Having bathed ourselves, we now want to make sure we stay fresh and so we may apply an antiperspirant or a deodorant. Thorough washing of the armpits can remove 99% of the bacteria but the remaining 1% can multiply rapidly and within a few hours will be happily feeding off sweat and releasing offensively smelling molecules. The human body has something like three million sweat glands and they are of two kinds: the eccrine and the apocrine. The former emits a 1% solution of sodium and potassium salts which does not smell, and its role is to cool the body by evaporation. The apocrine glands, on the other hand, which are found in the armpits,[16] groin and feet, emit sweat containing proteins, oils, steroids and cholesterol. This sweat is ideal for breeding bacteria like *Corynebacterium xerosis* and *Micrococcus luteus*, and these microbes break down the natural chemicals into the obnoxious compounds of which certain **carboxylic acids** are instantly recognisable and somewhat repulsive. There can be as many as ten million bacteria cells per square centimetre of armpit skin, and this is as true for both women and men, and yet the odour each sex emits is very different. Male underarm odour is much stronger and it has three components: an acrid smell which comes from the acids, of which 3-methyl-2-hexenoic acid is the most characteristic; a slight musky smell which comes from the steroids, and especially from androstenone; and a pungent smell which comes from sulfur-containing molecules, of which 3-methyl-3-sulfanylhexan-1-ol is particularly repulsive.

The bacteria which generate these odours can be tackled with antiperspirants and/or deodorants. Antiperspirants rely on aluminium and zirconium compounds for their action. These block the sweat glands by reacting with water to form a plug of aluminium hydroxide or zirconium hydroxide. When aluminium was suspected of causing Alzheimer's disease – wrongly as it turned out – manufacturers looked to zirconium as an alternative and this is still used in some formulations. Deodorants on the other hand are there to *kill* the bacteria and some people prefer to use these as well. A typical deodorant will contain around half a per cent of an antibacterial agent such as **triclosan** which can be incorporated into nanospheres so that it is released slowly during the day.

So having showered and deodorised you are now ready to face life in the modern, crowded, exciting city.

Chemistry is an essential part of urban living, saving both energy and water and will continue to do so to a greater extent this century. There will never be a sustainable city which provides all that its inhabitants require, but existing cities can be designed to consume less and waste less. While chemists can only play a minor role in bringing this about, their role is nevertheless a key one as this chapter has shown.

[16] Each armpit has around 25 000 sweat glands.

CHAPTER 7

Sport and Chemistry

[A word in **bold** means there is more information in the Glossary.]

Chemistry has transformed many sports but not always in ways that are sporting. The seven topics in this chapter are: sports equipment, sporting apparel, sports arenas, performance-enhancing drugs, performance-enhancing foods, Formula 1 racing and horse doping.

Sport is a force for good. It satisfies a human need to show physical prowess, to work as a team, to excel as an individual, and it can be exciting to watch. Some events such as the Olympic Games are seen by billions worldwide who enjoy this celebration of human endeavour and it is a wonderful way of bringing together young people from all countries and cultures. Despite its aims, there have been times when its high ideals have been subverted by politics, commerce, and unintentionally by chemists. Yet there is much that chemists have contributed to the world of sport which is beneficial and most of this chapter is about these positive inputs. Let us begin with a story of the 1800s when chemistry was just beginning to make an impact on society. In the sporting world this took the form of better snooker balls and those who benefited most were not humans, but elephants.

7.1 SPORTS EQUIPMENT

From better balls to better bikes, chemistry has delivered some remarkable results.

Snooker balls were originally made of wood, then of ceramics, and eventually of ivory, to the extent that by the 1850s around 20 000

A Healthy, Wealthy, Sustainable World
By John Emsley

Published by the Royal Society of Chemistry, www.rsc.org

elephants a year were hunted down and killed for their tusks alone, each one of which could only furnish enough of the natural polymer for 8 balls.[1] Ivory is dentin, a hard material consisting of calcium phosphate bonded to the protein collagen. The chemical alternative was a new plastic called celluloid which was made from cellulose nitrate. However, celluloid could be dangerously flammable, and there were tales that when the new billiard balls collided violently they sometimes went off with a bang. When this happened in a saloon in Colorado it started a gunfight, or so said John Hyatt, the man who developed celluloid. Today billiard balls are less exciting and are made of plastic, a resin made from aldehydes and phenol.

Golf balls, tennis balls and footballs have also benefited from chemistry. Golf balls were originally made of hard woods when the game was first played in Scotland in the 1400s, then in the 1600s they were made of leather stuffed with boiled goose feathers.[2] In the 1800s gutta-percha, a type of latex, was used and the balls were produced in moulds. Better still was a ball with a core of wound rubber thread which gave it much more resilience. Today's golf balls may have as many as four layers, with a core made from ethylene and acrylate copolymer. When these balls were first introduced they moved faster and went further but were rather hard, although they were later made less so by changing the copolymer blend. The core is surrounded by an outer layer of polyurethane, and covered with a softer copolymer of ethylene and propylene on which are stamped the characteristic dimples.

Footballs were once made of strips of leather sown together, inside of which was a rubber bladder that could be pumped up. They tended to get very heavy when they got wet. Now they have an outer shell of hard polyurethane with a layer of polyurethane foam beneath. These footballs are abrasion-resistant and unaffected by wet conditions. Their pentagon and hexagon panels are glued together to perfectly seal the ball. Even on a sodden surface such a ball picks up no water and its behaviour will not change during a match.

Ten-pin bowling is never likely to become an Olympic event but even its heavy bowling balls have improved. In 1960 they changed from being wood-based to ones made of rubber, and today they are made of polyester plastic. They consist of an inner core, an outer filler core, and an outer shell called the coverstock. The inner core is polyurethane, the outer filler core is either calcium carbonate or barium sulfate blended

[1] Ivory was also used for other things such as piano keys, cutlery handles, ornaments and trinkets.
[2] When these dried out they expanded to make a perfectly spherical ball.

with a resin, and the shell is polyurethane mixed with components that improve its grip on the well-oiled lanes.

Tennis balls have also improved. These can be either of the harder pressurised kind, favoured in North America, or the softer unpressurised kind favoured in Europe. The American ones tend to lose their bounce after a few matches whereas the European ones last longer but they deform momentarily when they are struck and that can affect play. In 2007 the chemical company Akzo Nobel introduced Micro X balls. These are filled with 700 million air-filled, plastic, micro-sized hollow spheres called Expancel, which are made from acrylonitrile-based copolymers. The International Tennis Federation, based in London, which certifies whether a particular brand of tennis ball can be used in championship matches, said that Micro X was acceptable.

Tennis racquets were once made of wood and to stop them warping they had to be tightly clamped in a press when not in play. Metal ones of steel or aluminium were free of this encumbrance and were lighter. Then came composites which were even better and finally we have the **carbon fibre** ones used by today's players which are even lighter and just as strong. The lighter the racquet, the more energy with which the player can hit the ball and it became possible to reach serves of 200 kph (125 mph). The first carbon fibre racquet to win at Wimbledon was that used by John McEnroe in 1983. Tennis racquets also have polyurethane strings which have the ability to withstand the force of powerful strokes.

Carbon fibres also play a part in cycling. Chris Boardman cycled to a gold medal in the Barcelona Olympics in 1992 when he established a new record in the 4000 metre Individual Pursuit. His bicycle relied very much on the benefits of carbon fibre. The top cycling event is the Tour de France which has 21 stages. The participants have to cover 3550 km (2200 miles) and different bicycles are chosen for different stages of the race. For the steep mountain stages the lighter the cycle the better, and carbon fibres give the best weight-to-stiffness combination. Almost as important is the aerodynamic design with disc wheels and aerodynamic handlebars – and with contestants wearing skin suits, tapered helmets and wraparound sunglasses. **Teflon** also has a part to play in bicycle lubricants which are used on chains and cables and are reputed to make pedalling easier and quieter, and make gear changing better.

Metal cycle frames are not entirely without merit and those constructed from new alloys made using scandium offer certain advantages. Scandium is not a rare metal but it is expensive to extract and purify, which explains why so little is produced and why it is costly.

Figure 7.1 When chemistry meets chemistry the serve can speed at 120 mph.

Nevertheless, it is worth doing because it can greatly strengthen other metals and it requires only a tiny amount of scandium added to aluminium to make a particularly strong aluminum alloy, strong enough to permit welding which ordinary aluminium will not. This new alloy has been used for baseball bats and for a short time it was used for cricket bats, but the latter were deemed unsporting and were forbidden.

Teflon plastic is the only surface to which a gecko lizard cannot cling, which gives you some idea of just how low its coefficient of friction really is, so it makes an ideal surface for boats and is included in antifouling paints. Teflon-containing waxes can be used on skis and snowboards thereby reducing their coefficient of friction. Coating the head grommet of a tennis racquet with Teflon is another way of making it more powerful. (The head grommet is the plastic strip with small tubes which is at the top of the racquet and through which the stings are threaded.) Normally the grommet would resist movement of the strings as they strike the ball whereas with Teflon this allows the strings to move without friction.

Kevlar is a polymer that has remarkable strength, flexibility, and is lightweight. When it is spun into fibres and heat-treated, the strands of polymer get even stronger, and they are used for fishing rods, tennis racquets, skis and running shoes. The plastic is five times stronger than steel and more elastic than carbon fibre, and it has pushed the performance limit of such equipment well beyond those of traditional materials. Golf clubs can have shafts of Kevlar making them lighter and so increasing the speed with which they come in contact with the ball. Kevlar is used for vaulting poles and javelins, which are now better than they have ever been, and also for canoes. A competition canoe made from Kevlar is so light it has to have ballast to bring its weight up to the minimum allowed, which is 9 kg.

7.2 SPORTING APPAREL

Man-made fibres help runners, riders and swimmers.

Clothing and shoes for sports people have been improving thanks to chemistry. One of the cardinal rules of running is never to wear cotton and the reason is that this fibre absorbs sweat and so it is likely to cause the runner to overheat. The answer is a garment made either from fine fibres of nylon or, better still, polyester. Neither type of fibre absorbs sweat and the weave can be so open as to allow this to evaporate.

The key to several kinds of sportswear is an elastic type of polyurethane invented by a DuPont chemist, Joseph Shivers, in 1959. Its generic name is elastane, but it is better known by its trade names of Spandex and Lycra. Spandex was ideal for corsets and girdles which were then popular items of female underwear, but it became more famous in the 1980s as Lycra. It could be blended with other fibres to make a stretchy textile which was unaffected by perspiration, body lotions and detergents, and was much better than natural rubber elastic being stronger, lighter and more flexible. It was ideal for swimsuits and bicycle shorts. Elastane will stretch to seven times its relaxed length and when the pressure is released it returns to its original state. Lycra vests and shorts are close-fitting which means less drag for runners and less risk of loose clothing knocking the bar off in a high jump. A tight-fitting Lycra outfit also reinforces the muscles and has been estimated to improve some performances by as much as 10%. When an athlete moves, then one set of muscles contracts while an opposite set extends and even this action can be improved by incorporating polyurethane bands in the apparel where it is in contact with the muscles. These function like springs and work in unison with the body's muscles.

Unlike the young men who took part in the original Olympic Games of ancient Greece and who appear to have dispensed with clothing altogether, the modern sports person of both sexes has to be clothed. Indeed in swimming, where clothing could be dispensed with, a body-hugging swimsuit can actually make a person swim faster. Unfortunately, chemistry's contribution to enhanced performance is too good and soon the wearing of such swimsuits for competition events will be banned.

Speedo developed Fastskin swimsuits for the Sydney Olympics in 2000 and the majority of medal winners wore them. The idea was that this knitted garment would mimic a shark's skin – and it worked. Then they developed Fastskin II for the Athens Olympics in 2004 and this had reduced drag. When Michael Phelps won six gold medals at the Athens Olympics in 2004 he wore a Fastskin swimsuit. The majority of medal winners in that event wore it.

Then Speedo launched an even better high-tech LZR Racer swimsuit in February 2008 which had 10% less drag, and within a few months its wearers clocked up more than 135 world records. The fabric of LZR Racer is composed of nylon-covered elastane. This not only makes it resistant to attack by the hypochlorite disinfectant of swimming pools but it creates a very tight weave. The suit is welded together, which dispenses with seams, and the swimsuit is finally exposed to a plasma finishing treatment to roughen the surface, but only at the nano level, and it then has a water repellent finish applied. Sadly all this has been to no long term benefit.

Fina is the world swimming governing body and it has banned all non-textile swimsuits and this ban will come into force in 2010 after the World Championships in Rome. Not all modern fabrics are affected. Nike's designs for Team GB are made from polyester and from recycled PET bottles. These garments were said to reduce drag by 7% which might benefit a runner by 0.02 second in the 100 m sprint. Not a lot, but maybe all that is needed to win.

Sports shoes are now much thinner than they used to be and they are highly elastic and flexible, with insoles of foam which make them excellent at absorbing shock. The foam is mainly a copolymer made of ethylene and vinyl acetate, although sometimes polyurethane foam is used. Some sports shoes contain Nike's Lunarlite foam which is super-light cushioning based on that designed by NASA for spacemen. It is made from a copolymer of ethylene, vinyl acetate and **nitrile rubber**. The soles of sports shoes are often made of polyurethane. Nike's Flywire sports shoes are remarkable in having paper thin soles and these have done away with most of the shoe upper, with high strength nylon fibres stitched across the top of the shoe supporting the foot just where it needs it most.

7.3 ARENAS

Underfoot and overhead, chemistry makes sport almost weather-proof, and for those who like to wallow in water, it makes that safer as well.

Artificial turf is now the surface on which many sports are played that were once played on grass. In 2006 more then 70 million square metres of artificial turf was installed worldwide and it's easy to see why: it doesn't become worn or pitted, or need watering, or cutting, or weed killers – and it can be used under cover, whereas grass needs the sun. The first artificial turf was Monsanto's AstroTurf which was introduced in 1966 and was installed in the Houston Astrodome. It was made from a polyamide polymer rather like nylon. It was tough but in fact was too tough, so that while it didn't wear out or break like grass, it was rather painful to fall on and it could cause abrasions. This first-generation turf was banned by the English Football Association in 1988 because it was too hard, was not springy like real turf, and balls would bounce off it too high and roll on the ground too quickly. Players who fell on it suffered joint injuries or grazes as they slid along it.

Since the mid-1990s, however, artificial turf has been making a comeback and now replicates the real thing very well. Instead of nylon fibres, artificial turf now has silicone-coated polyethylene fibres around 5 cm long which are softer than nylon and just as resilient. These are embedded in 4 cm of infill made up of sand and tiny rubber particles from old tyres. This also acts as a shock absorber making the impact of feet and bodies less damaging to both the turf and sportspeople. The fibres are sewn into a flexible and tough backing made of polyurethane, latex, or polypropylene. The newer artificial grass has clusters of four tough fibres which have a central spine, thereby keeping them upright, and four softer fibres that lie flat so that overall they behave very much like grass. The Dutch chemical company DSM has developed a new thermoplastic elastomer as a replacement for the rubber infill, and the Dow chemical company in the US has developed an elastomer-coated sand which can also be used.

Artificial turf can be laid onto concrete, or on to a layer of polyurethane foam which provides greater shock absorbency. Modern artificial turf behaves like the real thing in terms of balls bouncing and rolling, and players being able to run, slide and fall without injury. Moreover, it can be used all year round and for non-sporting purposes like rock concerts – and it needs no maintenance. This kind of turf is now acceptable to various sporting organisations, such as the global football association FIFA and the International Rugby Board, and it might even be used for the World Cup in 2010 in South Africa.

Despite their benefits, artificial pitches don't last forever and have to be replaced after 10 years. Then comes the issue of how recyclable they are, but even if this is not deemed possible they could be incinerated to generate electricity.

High-tech polycarbonate is the preferred material for the roofing of modern sports arenas because of its lightness and transparency. This was the material used for the stadium in Athens which was specially renovated for the 2004 Olympics.[3] At the time it was then the biggest stadium roof in the world and it is suspended on two metal arches. It provides not only protection against the rain but, and just as important, it is tinted to allow only visible light through so that the 75 000 spectators are protected from the suns' UV rays.

The centre court at Wimbledon was fitted with a retractable roof in 2009 which can be closed within 10 minutes if rain threatens, so that play will not be interrupted for any length of time. When the roof is closed, a large-capacity air-conditioning system ensures that players and spectators will not become over heated. The 7000 square metres of roofing material is a semi-transparent fabric made of **PTFE** fibres. It also allows natural light to reach the grass of the court.

Swimming pools are other arenas where chemistry is important, in this case in preventing swimmers being exposed to water-borne diseases such as typhoid, cholera, dysentery and meningitis. The disinfectant most used is sodium hypochlorite, which has the double benefit of being deadly to the disease pathogens and of persisting in the water for a long time – see Chapter 2. Other disinfectants like ozone are just as powerful and are used, but they have to be supplemented with hypochlorite because ozone does not persist. Even a level of 1 ppm (0.0001%) sodium hypochlorite kills 90% of germs, while at 5 ppm it will kill 99% of microbes within a few minutes.

---

[3] Named the "Spyridon Stadium" after the winner of the marathon in 1896.

Common Sense 16: The trouble with swimming pools is the strong smell of chlorine and the water contains organochlorine compounds which are dangerous

Wrong. It is not chlorine gas which is providing the smell but chloramine (NH_2Cl). This is produced when the hypochlorite disinfectant reacts with the urea in urine and that comes from swimmers, and especially from boys.

The organochlorines are formed in the water from traces of dissolved organic matter reacting with hypochlorite and there can be 20 times more organochlorines in pool water than in ordinary water. They originate from the organic components of human sweat and body oils. However, the levels are minute and unlikely to affect anyone. Nor does the formation of chloramine and organochlorines reduce the level of the disinfectant to any great extent; probably less than 1% of the hypochlorite is destroyed in forming them.

Suggestions have been made that swimming pools are responsible for asthma in some children and those who make this claim blame chlorine. However, there are so many confounding variables in the 'studies' on which such claims are based as to make them scientifically unsound. Were the use of chlorine to be ended then it might be that asthma cases would decline but we can be certain that many more children would catch waterborne diseases. In 2006 the World Health Organisation issued guidelines about the safety of places where people swim and bathe. Their concern was that as more and more people enjoy this type of recreation, more of them are being exposed to injuries and disease. As regards making the water safe, then its recommendations are to ensure the water is disinfected although not to such a level as to expose those swimming in it to other risks. The WHO says that chlorination at a level of 1 ppm is adequate to protect swimmers but could be half this dosage in a well maintained pool.

There are other ways of disinfecting swimming pools and other kinds of communal bathing water. **Trichloroisocyanuric acid** can be used and this is more stable when the pool is exposed to direct sunlight although it makes the water slightly acidic. Bromine was once used an alternative disinfectant for swimming pools when added in tiny amounts, as in the US in the 1930s, and it was less corrosive to equipment but more expensive. An alternative pool disinfectant which incorporates bromine is 1,3-dibromo-5,5-dimethylhydrantoin (DBDMH) and can be added to swimming pools when the recommended dosing is 10–50 grams per tonne of water.[4]

7.4 PERFORMANCE-ENHANCING DRUGS

Chemicals can improve sporting performance but those who use such drugs are always in danger of being found out.

In the last century sports people took amphetamines, **anabolic steroids**, hormones, painkillers and diuretics in order to improve their performance. For example, at the first of the modern Olympic Games, held in Athens in 1896, the 100 metres race was won by Thomas Burke of the US in 12 seconds. When Ben Johnson achieved a gold medal for winning this race at the Seoul Olympics in 1988 his record time was 9.7 seconds. He was then disqualified because he had taken Stanozolol which was detected in his urine. Stanozolol promotes muscle growth without causing weight gain. In the 2008 Olympics in Beijing the record was broken yet again when Usain Bolt of Jamaica recorded a time of 9.69 seconds, and without being disqualified.

Performance-enhancing drugs have been a problem in the sporting world for more than 50 years. There are various classes of prohibited substances and the World Anti-Doping Association (WADA) tests for ones that might be used. Stimulants like **caffeine**, amphetamines and ephedrine increase alertness and even induce a feeling of aggression, and have been used in long-distance running, cycling, American football and baseball. Narcotics like heroin and methadone reduce sensitivity to pain and have been used in boxing and contact sports. Anabolic steroids like testosterone, nandrolone and tetrahydrogestrinone are used in all sports and especially weight-lifting, gymnastics and track events. Diuretics like caffeine and **mannitol** promote rapid weight loss *via* urine and are used in boxing and wrestling, where bodyweight determines the contest category. Hormones such as human growth hormone and erythropoietin are said to increase muscle strength and endurance, although this remains unproven, and they have been used in various sports where strength really matters, such as cycling and long-distance running.

Caffeine can unlock the body's supply of stored energy, and it stimulates the release of calcium into muscle tissue, both of which make a person more active. Those taking part in cycle races drink caffeine both before and during the race, and some even insert caffeine suppositories to release the drug slowly. But caffeine can exact a high price,

[4] A tonne of water is 1000 litres.

as Sylvia Gerasch discovered. She was the European 100-metre breast stroke champion, but she tested positive for the drug and was duly stripped of her title and banned from competitive events for two years. That was in January 1994, when she was found to have 16 mg per litre of caffeine in her blood, which was over the limit of 12 mg permitted by the authorities. This level could be achieved by drinking lots of strong coffee in the half-hour before an event, but to reach the level found in Gerasch she would have had to take in around 750 mg of the drug, which would be like drinking 20 cups of coffee.

Another early performance enhancer designed to keep a person alert was benzedrine which was used by football teams in the 1960s. (It was given to bomber crews in World War II when they were on long flights.) Cyclists also took these pills.

Neither caffeine nor benzedrine is to be compared with the performance-enhancing products many athletes chose to take in the second half of the last century. The most popular were the anabolic steroids. These were first isolated in the 1930s and used as a medical treatment to stimulate puberty in boys in whom it was delayed and to aid growth. But it was not until the 1950s that sportsmen began to use them, especially when it was desirable to increase body mass. The list of those caught trying to gain an unfair advantage with these drugs contains some famous names, including those who have won medals at the Olympics. Anabolic steroids have turned up in athletics, baseball, marathons, boxing, football, wrestling, cycling and weight-lifting, and in the last of these it was even detected in athletes at the prestigious Highland Games in Scotland in 2009. Floyd Landis who won the 2006 Tour de France lost his title because he has used them.

Weightlifters began using testosterone in the 1950s as a way to build up muscles, generally at the expense of those organs of the body which produce it naturally so their testicles became redundant and generally shrank in size, a further indication of what they were doing. Then, in 1958, a synthetic anabolic steroid methandrostenolone (aka Dianabol[5]) became available which did the job even better and without side effects and that's what Arnold Schwarzenegger injected. Soon other athletes were using it and similar drugs, such as Turinabol, which is testosterone with a chlorine atom added. This worked even better in that it could be taken orally, and it was metabolised quickly so it escaped detection by the methods of drug testing then available.

---

[5] Dianabol is no longer produced.

The former East German Government basked in the welter of Olympic medals which its teams won in 1968, 1972 and 1976. The total was truly impressive for a country of its size – population 17 million. At the Mexico Olympics of 1968 they won 9 gold medals, at the 1972 Olympics in Munich they won 20, and at the 1976 Olympics in Montreal they won an unbelievable 40. The fact that some of their top women athletes looked and spoke like men did not go unnoticed, however. The IOC were aware of what was going on and by the early 1980s they had found a way of proving that an athlete was taking testosterone-like drugs. Too much testosterone meant that the athlete was disqualified and several met this fate. However, the East German chemists were one step ahead and found a testosterone-like drug which the body could rapidly eliminate. If an athlete stopped taking this a few days before a competition they would not fail the urine test. Following the re-unification of German in the early 1990s, the organisers of the former East German Olympic programme were brought to trial and found guilty of inflicting actual bodily harm to around 100 young women by giving them anabolic steroids.

Not that any of this has stopped athletes from taking anabolic steroids. The Ukrainian heptathlete Ludmila Blonska tested positive for Stanozolol in 2003 and had a two-year ban imposed on her. She also lost an Olympic silver medal in 2008 when she tested positive for another anabolic steroid, methyltestosterone. One of the most famous cheats was Ben Johnson, the sprinter who set several world records. His most prestigous win was the 100 metres at the 1988 Olympics, when he beat his main rival Carl Lewis by 0.06 second at 9.93 seconds. A urine sample taken after the race proved positive for Stanozolol and he was disqualified, and he later admitted taking steroids saying he had been doing so since 1981. When Ben Johnson was stripped of his Olympic gold medal it automatically promoted Carl Lewis to first place despite the fact that, prior to these games, he had been found guilty of using a banned substance. He successfully argued that this had been part of some medicine which he was taking at the time. In 1993 at a race in Montreal, Ben Johnson again tested positive, this time having injected testosterone, and was banned for life. All he could now do was run alone and try and outdo his previous performances but even when he did this in 1999 he was caught using another drug, hydrochlorothiazide.

Stanozolol is an anabolic steroid derived from testosterone, and available as the suggestively named Winstrol. It is approved for medical use and can be taken in tablet form to treat anaemia and other conditions. It stimulates the production of red blood cells, enhances muscle growth and increases bone density. Clearly these are advantageous to an athlete, which is why so many take it. However, for female athletes it can give them male characteristics.

A safer anabolic steroid is Furazabol. Its growth-enhancing effects are the same as Stanozolol but it also lowers cholesterol levels. Another anabolic steroid is Nandralone, which is also present naturally in the human body but if it is used then the level far exceeds the natural level. That was how the British sprinter Linford Christie was found out, as was the Pakistani Test cricketer Shoaib Akhtar, and footballers Edgar Davids and Christophe Dugarry.

The man who led the fight of analytical chemists against the drug takers was Don Catlin, who in 1982 set up the Olympic Analytical Laboratory in California. He used chemical analysis to demonstrate the use of drugs. Today there are powerful methods of doing this such as gas chromatography (GC) or liquid chromatography (LC) linked to mass spectrometry (MS). With **GC-MS** or **LC-MS** it is now possible to identify even the tiniest amounts of substances in blood and urine.

Common Sense 17: Some performance-enhancing drugs are undetectable

Wrong. At one time it was possible to take hydrochlorothiazide which would mask the presence of other drugs, but its presence was of course suspicious. Analytical chemists are now able to detect traces of drugs which an athlete might have ceased using several months ago. They can even prove that an athlete has used chemicals which are indistinguishable from those which occur naturally in the body such as testosterone. To do this analytical chemists measure the carbon-13 : carbon-12 ratio. There is more carbon-13 in human testosterone than in the testosterone manufactured from soya, which is where most of the artificial testosterone comes from.

Forensic chemists concentrate on finding traces of the drug itself or the metabolite into which the drug has chemically changed while in the body. The analyst first converts these to their volatile derivatives by attaching a trimethylsilyl group [$(CH_3)_3Si$] and this enables GC-MS to identify them exactly.

A few sports persons have gone to great length to disguise their drug taking. Some have even hidden a sample of clean urine to submit for testing, and in extreme cases this has even been done by filling their own bladder with it using a catheter, or in the case of men by wearing a false penis from which the urine could be seen to be delivered, although this was really from a hidden pouch attached to their body. A slightly less dramatic method is to take diuretics to increase urine output before having to provide a sample, although this might not remove all that the person is trying to hide.

The Bay Area Laboratory Cooperative (BALCO)[6] was the source of many anabolic steroids and its products were designed to be one jump ahead of the analytical chemists. The company even produced one performance-enhancing drug which did not show up in the forensic tests – and that was tetrahydrogestrinone (referred to as 'The Clear' by its users). Eventually someone informed the US Anti-Doping Agency of what was going on and named Victor Conte of BALCO as the source of this new drug. The whistle-blower also sent a used syringe containing traces of The Clear which had been retrieved from a waste bin.

The US Anti-Doping chemists now knew what they had been missing, so they devoted their efforts to find a way of detecting it even though it slipped through the conventional GC-MS analysis method, its signal being indistinguishable from those of natural testosterone. After a great deal of effort, the chemists identified the traces of material in the syringe as tetrahydrogestrinone and then made some themselves. They tested its effect on a baboon and were able to show that there were differences in the animal's urine that they could observe. Small though these differences were, it finally enabled the users of The Clear to be identified. When US Police raided BALCO they found lists of customers including Olympic champions and a famous baseball star. Victor Conte was sent to prison for four months in 2005 after being found guilty of supplying illegal drugs.

In 2003 four champions in the American National Track and Field Championships, and later a British sprinter Dwaine Chambers, were found to be using The Clear. Chambers was given a two-year ban, but he went on to win a gold medal in the European Championships in August 2006. In January 2008 Marion Jones, the American track athlete, pleaded guilty to lying to US federal prosecutors about her use of the drug and she was sentenced to six months in prison and stripped of the five medals she had won at the 2000 Sydney Olympics.

The most difficult drugs to test for are now erythropoietin (EPO) and human growth hormone (hGH). The former is prescribed legitimately to treat anaemia because it encourages the growth of red blood cells thereby giving blood a greater capacity to carry oxygen, which clearly can be of help to athletes. The latter is a complex molecule with

[6] Based at Burlingame, near San Francisco airport.

carbohydrate and protein components, and it is the hormone which controls the body's production of red blood cells and it is produced by the kidneys. It is also involved in wound healing. EPO can be taken as a performance-enhancing drug and it can be detected because there is a slight difference between its carbohydrate component and that which occurs naturally in the human body.

hGH is a hormone which stimulates growth. Although all its functions have still to be identified, it is known to increase muscles and bone density while decreasing body fat, hence its attraction to athletes and bodybuilders. Its use was banned by the IOC in 1989, but had been widely used in the 1970s, 80s and 90s until it was proved in the early years of this century that artificial hGH could be distinguished in blood samples. Research into finding ways of detecting this kind of abuse is now being undertaken, including ways of proving that someone has had a massive blood transfusion prior to a sporting event.

Immunoassay is another method of testing for banned substances when these have been given in doses almost too minute to be detected by normal chemical analysis. Biomarker-based testing will almost certainly be available for the London 2012 Olympics and this will enable drugs to be detected after their user has ceased using them. This is particularly important for those substances which are quickly excreted from the body such as hGH. Two protein biomarkers have now been identified which are raised by hGH and which remain high for weeks afterwards.

Immunoassay is based on the immune system's ability to recognise alien molecules (antigens) and once it has done this it produces a molecule called an antibody that bonds specifically to the antigen. In this way, the antibody identifies the antigen as alien and leaves it to be destroyed by other parts of the immune system. The antibody itself betrays the former presence of the drug and there is now a range of antibodies that are specific to particular drug molecules or their metabolites, which are the molecule that the liver converts the drug molecules into. Immunoassay permits a urine sample to be analysed without lengthy preparations and it is cheap and quick.

The media highlight athletes who use performance-enhancing drugs, but we should keep this in perspective. According to the WADA, at the 2004 Athens Olympics only 26 positives were detected, while at the 2008 Beijing Olympics only 6 positives were recorded. Similar low figures are reported by the US Anti-Doping Agency on the many tests they carry out each year, and the World Anti-Doping Agency, which tested 25 000 samples in 2007, found that only 2% tested positive.

7.5 PERFORMANCE-ENHANCING FOODS

There are chemicals in foods which provide legitimate ways of boosting sporting performance.

It is possible for athletes to boost their performance by taking perfectly acceptable chemical substances. Food supplements and energy drinks are there to help. Muscle needs protein for its structure and glucose to provide energy, and these require a diet rich in amino acids and digestible carbohydrate both of which can be obtained from food.

Energy comes from glucose and it is stored in every cell of the body as the polymer glycogen. This supplies the energy not directly, but by producing ATP (adenosine triphosphate) which is the real energy driver. Clearly the more glycogen the body can accumulate the more fuel it has to call upon, so an athlete should eat lots of carbohydrate in the days before the event. This 'carbohydrate loading' will be achieved by eating foods like rice, bread, pasta and potatoes.

To maintain a constant level of glucose in the blood, the liver either converts it to glycogen if there is a surplus or releases it from glycogen if it is needed. There can be up to 1 kg of available glycogen in the body,

Figure 7.2 What's the chemistry in a sports drink that can boost your performance?

capable of yielding as much as 4000 food Calories of energy. When the glycogen has been used up, then muscles will cry out for rest and that is likely to happen when a person has engaged in strenuous exercise for an hour or so.

Glucose can release its energy in one of two ways: by reacting it with oxygen to form carbon dioxide and that is called 'aerobic' respiration, or if there is insufficient oxygen around it can convert it to lactic acid and that is called 'anaerobic' respiration. Normally we rely on aerobic respiration, but an athlete's body makes extreme demands for energy so anaerobic respiration soon kicks in, popularly referred to as getting your 'second wind'. A build-up of lactic acid in the muscles will eventually result in a feeling of fatigue, although it can be countered by 'bicarbonate loading' which means taking large doses of sodium bicarbonate ($NaHCO_3$) just before an event. This makes the blood more alkaline so that it can neutralise and solubilise the lactic acid and thereby remove it more quickly from the muscles.

It is possible during training to measure the concentration of lactic acid in the blood using a test strip and a drop of blood from the finger tip. Knowing this can help athletes to maximise their performance. The normal level of lactic acid in blood is less than 0.1 g per litre but it can rise to 2.25 g per litre after a 400 metre sprint. As the heart begins to beat faster, so the level of lactic acid rises, and markedly so in an endurance event. Then an athlete has to keep the heart rate down to a level which does not interfere with performance. Training can increase the rate at which the heart can beat without lactic acid interfering.

A marathon runner who completes the course in 3 hours will consume as much energy as a 1 kW electric heater uses in an hour. Although he or she generates a lot of heat in the process, the body stops itself from overheating by evaporating about two litres of sweat from the skin. Along with the sweat, the body also loses up to four grams of sodium and about half a gram of potassium. Then an energy drink is called for, and this will mainly contain glucose, which the body can absorb quickly and use immediately. A typical sport drink will have around 60 grams of glucose per litre. It will also contain sodium and potassium. In the UK we have Lucozade Sport and in the USA there is Gatorade, which is now also available in the UK. Lucozade Sport contains 64 grams of carbohydrate, 1 gram of sodium, and 100 milligrams of potassium per litre while Gatorade has 56 grams of carbohydrate, 4.5 grams of sodium, and 120 milligrams of potassium.

Other chemicals which sports people need to obtain from their diet or from supplements are: amino acids, alanine, creatine, growth hormone stimulators, vitamin and mineral supplements, nitric oxide precursors and protein. Professional athletes should take advice from a qualified dietitian but for most sports enthusiasts there are numerous fitness supplements they can take. These will often include creatine. Creatine comes with eating fish and meat, although it can be synthesised naturally in the kidneys from the amino acids glycine and arginine. Arginine is an amino acid which is not normally regarded as being an essential part of the diet because the body can make it. However, the body doesn't produce it in sufficient quantities and so some arginine must also come from the food we eat. Arginine is synthesised from citrulline which is itself made from glutamine and this is done in the small bowel. Arginine is needed by the body to make nitric oxide which increases blood flow by relaxing blood vessels. Arginine also stimulates the release of growth hormone thereby increasing muscle mass and at the same time reducing body fat.

Another supplement is glutamine, which again is a natural chemical and it is said to reduce fatigue, build muscle and boost the immune system. Tryptophan can also be taken and this is said to boost adrenalin which is useful at raising the body's performance. Such dietary supplements can ensure that an athlete is not underperforming because of some dietary deficiency. Then there are products with enticing names like NO.Xplode, Shotgun Powder, Super Dymatize and the like, which claim to provide energy and help recovery during events, and some say they can make the body stronger and fitter for action.

After he had served his four months in prison, Victor Conte, the man who supplied tetrahydrogestrinone, set up a new company Scientific Nutrition for Advanced Conditioning (SNAC) selling performance-enhancing foods. These don't rely on sexy names but are marketed under more scientific-sounding brands as Aerobitine, Proglycosyn and ZMA. What do they contain?

Aerobitine is supposed to reduce fatigue and promote stamina, strength and endurance which it does by slowing the build-up of lactic acid and increasing the nitric oxide levels in the tissue which improves blood circulation – or at least that is what Conte claims. In fact Aerobitine is a mixture of vitamins C and E, folic acid, selenium, and the amino acid arginine, which increases the production of the growth hormone, and citrulline, which is an amino acid normally present in protein. Citrulline can reduce muscle fatigue and relaxes blood vessels. It is a natural chemical and is also present in the rind of watermelon.

Proglycosyn is an orange-flavoured mixture of protein and glycogen, along with dextrose and various minerals (in particular magnesium, calcium and **chromium**, as well as sodium and potassium), and it appears to be derived from whey. Proglycosyn is supposed to help the body recover after a workout.

ZMA is short for zinc and magnesium aspartate and it comes with vitamin B_6 (pyridoxine) and is meant to replace these minerals and this vitamin which are lost during exercise. Vitamin B_6 is involved in red blood cell production as well as regulating the balance between sodium and potassium, which is how the nervous system conveys messages from the brain to muscles.

Chemistry clearly has a part to play in performance-enhancing foods and food chemists must be involved in devising these products. When it comes to our next topic then the physical prowess of the sportsman – and they appear all to be men – is less important than their mental prowess. Aggression and risk-taking are what drives them to win while they push to extremes a machine in which chemistry plays a vital role. We are talking about Formula 1 Grand Prix racing.

7.6 FORMULA 1 (F1)

Although it appears to be mainly a sport linked to oil companies and car manufacturers, F1 racing owes a lot to the chemical industries.

Chemicals can give an unfair advantage to some F1 cars. At the Brazilian Grand Prix in 1995 Michael Schumacher came first and David Coulthard was second. Unfortunately, their F1 racing cars had been running on fuels that were contrary to new regulations laid down by the Fédération Internationale de l'Automobile (FIA). Both teams were fined $200 000 and both drivers lost championship points. Adding chemical boosters to the petrol tank was no longer admissible, although it had been allowed until the early 1990s. In fact the FIA would like Shell, Elf and Mobil who supply fuel for F1 cars, to provide something that is not too different from that which is available to motorists in general. All teams are now required to provide samples of fuel for analysis ahead of the race.

When a fuel is under suspicion then a sample is taken and is divided into three parts: one held by the team, one analysed at the track and one retained by the FIA in case of dispute. In 1997 Mika Häkkinen's team was found to have tampered with the fuel and the FIA stripped him of his third place at the Belgium Grand Prix and fined the McLaren team $50 000 (£30 000).[7] They appealed, so the third sample was then sent to

[7] Michael Schumacher now came first, Giancarlo Fisichella was second, and Heinz-Harald Frentzen was third.

Figure 7.3 A Formula 1 motor depends for its success on the chemical formula of its components.

an independent testing laboratory in Germany whose results supported the FIA findings.

The fuel for F1 cars has to meet strictly defined guidelines and its various components are spelt out in the form of the PONA list; P stands for paraffins, O is for olefins, N is for naphthenes and A is for aromatics. The FIA sets a maximum percentage of each category that is permitted, although it allows 5.75% of the fuel to be oxygenated molecules, such as alcohols, like butanol, or ethers, like diisopropyl ether, because these promote cleaner burning within the engine.

F1 racing is at the cutting edge of engineering technology and chemistry makes a contribution when it comes to bodywork, lubricating oil, tyres, brakes, batteries and driver protection. The bodywork has to be strong and light; the oil has to protect the engine under extreme stress; the tyres have to maximise grip under varying road conditions; the brakes will run red hot; the battery has to be as light as possible; and the driver has to be able to walk away from a crash unharmed.

Bodywork. High-speed cornering and severe braking put the car under great stress. The bodywork has to provide the strength to cope with this and yet be as light as possible, so carbon fibre-reinforced plastics are used. However, the minimum weight of car + fuel + driver must be at least 605 kg. Some cars weigh as little as 450 kg, so ballast has to be added although this can be an advantage in providing better weight distribution.

Lubrication. Chemistry has an important role to play in lubrication in all motors, and thanks to chemistry the oil in a normal car need only be changed after 30 000 km (20 000 miles). The oil in a F1 engine needs replacing every 400 km (250 miles). Lubricating oil is 80% base oil consisting of long chain hydrocarbons, with between 20 and 50 carbon atoms, and obtained from crude oil. To this are added detergents, friction modifiers, viscosity improvers and anti-wear compounds. These prevent engine wear and keep it clean, and they ensure that the oil continues to work at low and high temperatures, which it does thanks to the viscosity improver. This will be a polymer like poly(methyl methacrylate) or poly(ethylene–propylene). As the temperature rises these uncurl and in so doing they keep the viscosity within its working range. Additives like sodium phenoxides act to neutralise the acids which form during combustion and they also act to keep soot particles suspended so these can be filtered out. The anti-wear additive is **zinc dialkyldithiophosphate** and this coats the metal surface of the engine with a micrometer-thick layer which protects against abrasion and rust.

Tyres. Whereas the tyres on the average car consist of rubber reinforced with steel radial plies, those on an F1 car are made with soft rubber and with nylon and polyester plies which are designed to withstand much larger forces and to grip the surface of the track. The softness is achieved by using less sulfur when vulcanising the rubber and by the addition of oil, and the more of this, the softer will be the tyre. In 2009 the FIA allowed so-called 'slick' tyres to be used again and these have no tread so they provide better contact with the track. The disadvantage is that soft tyres are easily torn and only survive a few laps before needing to be changed. Moreover, the tyres have to be warmed up to optimise their performance. During the race the tyres will reach temperatures of 80–110 °C and they can only be inflated with air, nitrogen or CO_2.

Brakes. An F1 car can brake from 200 kph (120 mph) to a complete stop in less than 3 seconds during which time it travels only 65 metres.[8] Skilful braking is all part of the race and the brakes play a vital role. Unlike those of a family car, the brakes on a F1 can run red hot and for

[8] Acceleration is even more remarkable with 0 to 100 kph (60 mph) in less than 2 seconds, and 0 to 300 kph (180 mph) in less than 9 seconds.

this reason they are made of ceramic or carbon fibre. Carbon composites were first used by the Brabham team in 1976 and they are able to work up to 1000 °C.

A braking car is wasting energy, but that can be reclaimed to some extent using a regenerative brake. This is one which captures the energy of braking and stores it for future use and this can be done either by converting it to electrical energy and charging an ultracapacitor, or by converting it to the motion of a flywheel. The FIA started allowing the use of 60 kW kinetic energy recovery systems (aka KERS) in the 2009 season with either system permitted.

An electric double-layer capacitor, a so-called ultracapacitor, can absorb an enormous amount of electrical energy in an instant, and release it quickly as well. It consists of two layers of carbon separated by a thin porous insulator and sandwiched between two electrodes. Most contain carbon in the form of activated charcoal, **graphene**, or carbon nanotubes.[9] Yet other materials are being researched by chemists. The ultracapacitor market already exceeds $400 million a year and is growing rapidly. There are even buses in China which store their electricity in this form and which recharge at every bus stop.

Ultracapacitors can be as small as a conventional battery and yet within them is this special material which can have a surface area of a 2000 square metres per gram.[10] They can store up to 20 times more energy than a battery and they can go through the charge/discharge cycle millions of times without it affecting performance whereas a lead–acid battery begins to fail after it has been through this cycle around 1000 times. Ultracapacitors continue to perform well even at temperatures as low as −40 °C. However, in F1 cars the battery is the dominant source of electricity.

The alternative KERS is that of the Flybrid type which relies on a flywheel. This looks rather like a drum brake and weighs 25 kg with an energy capacity of 400 kJ. Its 24 cm diameter flywheel weighs 5 kg and inside its vacuum chamber it can revolve at more than 60 000 rpm. This form of energy retrieval and storage is already being used on railway engines. The company Flybrid has demonstrated a gyro made of steel and carbon fibre suitable for F1 racing cars.

Battery. F1 racing cars by their very nature don't want the kind of lead–acid battery of the family car. Their battery need only be small and **lithium batteries** seem ideal though these are potentially dangerous if

[9] Graphene is a sheet of carbon atoms one layer thick. (Graphite itself consists of stacked layers of such sheets.) Nanotubes are narrow sheets of graphene rolled into tubes.
[10] That's an area equivalent to around 1/4 of a football pitch.

they are damaged and the lithium metal starts to burn. Toyota in Japan has already developed lithium ion batteries suitable for cars and these are the lightest and most efficient kind of battery. Until recently they were only used in small personal items like cameras and laptops.

Driver Protection. F1 racing is a dangerous sport. Protecting the driver is done in the form of a lightweight, fibre-reinforced plastic helmet which can withstand severe impact. The clothing he wears – suits, balaclava, gloves, long johns and socks – must also be comfortable yet fire resistant and the preferred material is Nomex, which is a Kevlar-type polymer widely used by fire-fighters. Kevlar's remarkable strength is also incorporated into the laminate used to reinforce the driver's survival cell.

F1 racing cars have been test beds for all kinds of innovations. The sport continues to excite audiences like few other sports, and while it remains at the forefront of engineering and technical advances, its existence is assured. Meanwhile chemistry will be there to play its part.

7.7 HORSE DOPING

Another sport where chemistry has been misused and where chemists are now preventing it.

Horse races are in a different league financially when compared to other kinds of races in that millions of pounds may be gambled on their outcome, and that may be determined not by the animal's ability but by the substances it has been given. To prevent this happening there is HFL Sport Science (now part of Quotient Bioresearch, formerly known as the Horseracing Forensic Laboratory) at Newmarket which is committed to stamping out horse doping. Happily this crime is not common and less than 0.3% of horses tested in the UK are found to have prohibited substances in their system.

Doping horses does occur, and at the Beijing Olympics a group of riders from the show jumping teams of Ireland, Germany, Norway and Brazil were disqualified because their horses has traces of capsaicin in their urine. This is a natural chemical – it gives chillies their hot flavour – and it can act as a painkiller and for this reason it is banned. In May 2009 one of the Queen's horses, Moonlit Path, was found to have tranexamic acid in its urine. Tranexamic acid is a pharmacological drug used to reduce bleeding but one which also acts as a performance booster. The trainer said that the drug had been administered by a vet for health reasons.

Figure 7.4 The punters need to be assured that a race is fairly won, and chemical analysis can prove that it is – or it isn't.

Animals can bc given drugs to calm them down prior to being transported by air and the one that is given is acepromazine. However, once the horse has arrived at the race meeting it has to be given time for this to be metabolised out of its system before it takes part in a race. A horse can also be doped to make it appear healthier when it is put out to stud, or is being sold.

Horse doping was clearly a problem in England as long ago as the 1600s, when a regulation of 1666 banned the use of 'exciting' substances at Worksop races. However, it was really in the USA that horse doping became endemic and the same thing began to happen in Europe in the first part of the 1900s. In the US it was the FBI which cleaned up the sport with raids, arrests and trials, and the saliva of horses was tested to prove that doping had occurred. After World War II, doping gangs operated in the UK and several of their members were brought to trial.

Chemicals given to horses can either make them appear better than they really are or to make them run less well than they are capable of doing. Among the former types of drugs are procaine and oxyphenbutazone used as painkillers, and caffeine, theobromine and amphetamines used as stimulants, and the object is to give a less fancied horse a better chance of winning. Among the drugs which can slow down a horse are things like phenylphenidate and morphine, and here the objective is to reduce the chances of a favourite winning while placing bets on the second favourite.

Caffeine is a stimulant and while in the form of tea, coffee and colas it has only a modest effect on humans, it has a much greater effect on horses. Theobromine is very similar to caffeine except it is smaller by one methyl group, but it is just as potent nevertheless. These natural chemicals are part of chocolate which is how they could easily be administered to a horse and why the race-going public has to be kept away from horses before a race.

Other drugs which have been used to dope horses are flunixin, promazine, phenylbutazone and methylprednisolone. Flunixin is a powerful painkiller and is a non-steroidal, antiinflammatory drug used by vets to treat horses suffering from muscular pain and joint disease. Flunixin is given in conjunction with other drugs to protect the horse's digestive system because it can be very irritating to the stomach and gut. Flunixin is a prohibited drug under racing rules and it can be detected in the urine for as long as two weeks after administration.

Promazine is a tranquilising drug used legitimately to treat dogs, cats and horses. It relaxes the central nervous system and thereby calms the animal. Its advantage as a drug is that is works within 45 minutes of oral administration and there is no trace of it in the body after three days.

Painkillers can make a horse perform better and the ones typically used are phenylbutazone (used to treat osteoporosis and rheumatism) and methylprednisolone (an anabolic steroid type of drug that will reduce tissue inflammation and suppress the immune system).

What really helped stamp out horse doping was the introduction of modern methods of urine analysis which began in the 1950s. Routine dope testing started in 1963 and around that time there were three major conspiracy trials in the UK which ended in convictions and long prison sentences. One gang specialised in secretly gaining access to a racehorse and feeding it large amounts of caffeine in the hope that it would win.

As with human tests, a sample of urine or blood is taken from the horse and split into A and B samples. The A sample is used in the analysis with the B sample being kept in a deep freeze. If the test turns out to be negative then both samples are destroyed. However, if it turns out to be positive then the B sample is sent to another laboratory for confirmation, and if this comes back positive as well then the horse may have been doped and an enquiry is launched. (Of course there is also the possibility that the horse has been given medication.)

When the stewards at a race meeting decide that something looks suspicious, such as an outsider winning or the favourite performing poorly, then they will take a sample of that horse's urine or blood. Apparently it is relatively easy to get a horse to urinate; the noise of rustling straw will almost always trigger this. Urine used to be preferred for analysis because its composition does not change, but blood is now more widely used. The sample is then sent to HFL Sport Science and is analysed using a variety of techniques including GC-MS and without its identify being revealed to the analysts.

Steroids are metabolised as glucuronide conjugates. Adding enzyme breaks these conjugate bonds, leaving the free drug available for detection during analysis. The sample is then passed though a silica cartridge which absorbs the drug molecules but lets the other soluble components pass through. The silica is then washed with various liquids to remove any interfering material before it is washed with solvents to remove the drug molecules, which then go on to be separated by GC or LC and identified by MS.

Immunoassay is also used and HFL Sport Science has a range of antibodies that are specific to particular drug molecules or their metabolites, which are the molecules which the liver converts the drug molecules into. Immunoassay permits a urine sample to be analysed without lengthy preparation and it is cheap and quick to use.

Sport appears to be an essential part of life and chemistry is now an essential part of sport. Can it remain so in a sustainable future? I believe it can. There is nothing mentioned in the section on equipment, apparel, arenas and cars which could not be made from sustainable resources.

Something to Think About

What can we expect in 7, 17 and 27 years time?

At the millennium, newspapers delighted in looking back 100 years and seeing what kind of future the Victorians were predicting in 1900 for the century which lay ahead. It turned out to be nothing like what they imagined, not least in terms of scientific discoveries. In fact it would seem that whatever is predicted for the future will invariably turn out to be wrong. So why should I add to the list of failed soothsayers? And even if I were to be correct, how can I know that I won't suffer the same fate as Cassandra of Troy? The gods gave her the power to foresee the future along with the curse that what she forecast would not be believed. I'm fairly certain that the future on which this book is based, *i.e.* that there will be a sustainable world, will come about, if not this century then the next. The alternative, of slow decline to a pre-industrial society, would be a betrayal of future generations.

Nothing daunted, I have decided to make some limited predictions derived from the themes of this book and projecting them seven, seventeen and twenty seven years into the future. Why those years? Seven of course is a rather special number when it comes to predicting, witness the predictions Joseph made for Pharaoh of seven years of plentiful crops to be followed by seven years of famine. It appears he got things right. Seven, seventeen and twenty seven years are almost within the compass of what remains of my own life, so in making them I might have to live with the possibility of seeing them be proved wrong or maybe proved right. Here goes:

A Healthy, Wealthy, Sustainable World
By John Emsley

Published by the Royal Society of Chemistry, www.rsc.org

1 FOOD AND CHEMISTRY

In 7 years: GM crops will be generally accepted even in Europe. There will be better targeted chemical pesticides; microchip monitors will check food all the way from farm to processing plant to warehouse and finally to the supermarket.

In 17 years: There will be small-scale, sustainable, low-energy, nitrogen fertiliser plants around the world.

In 27 years: Most back gardens will grow GM bushes with high energy fruits enabling individuals to produce a meaningful amount of their own food.

2 WATER AND CHEMISTRY

In 7 years: A safe supply of drinking water for almost everyone on Earth.

In 17 years: Automatic water monitoring of water supplies at all stages.

In 27 years: Water-efficient homes whose roofs will collect water which can then be disinfected electrically to a level of cleanliness suitable for most household use.

3 HEALING DRUGS AND CHEMISTRY

In 7 years: Better drugs to treat viral diseases and a cure for arthritis.

In 17 years: Gene testing of individuals will enable drugs to be prescribed which will be better targeted and not produce side-effects.

In 27 years: Early detection of cancers, plus drugs to destroy them. One woman alive at this time will be 125 and a man will be 120.

4 BIOFUELS AND CHEMISTRY

In 7 years: 10% of motor fuel will be biofuel.

In 17 years: 25% of motor fuel will be biofuel.

In 27 years: Aircraft will be fuelled by liquid hydrogen.

5 PLASTICS AND CHEMISTRY

In 7 years: 5% of plastics will be sustainable bioplastics.

In 17 years: 20% of plastics will be sustainable bioplastics.

In 27 years: 50% of plastics will be sustainable bioplastics.

6 CITIES AND CHEMISTRY

In 7 years: Street lighting and indeed most outdoor lighting will rely mainly on solar panels.
In 17 years: Cities will generate 10% of their energy and use 10% less water.
In 27 years: Cities will generate 25% of their energy and use 25% less water.

7 SPORT AND CHEMISTRY

2016 Olympics: No contestant will be found to have taken drugs.
2020 Olympics: All athletes will be monitored non-invasively at the start of events.
2024 Olympics: These will be held entirely undercover in Saudi Arabia or Dubai.

Glossary

A word in ***bold italics*** means that there is an entry in the Glossary under that word also.

Acrylonitrile (aka 2-propenenitrile) has the chemical formula $CH_2{=}CHCN$ and is the liquid precursor of polymers such as acrylic fibres (trade name: Orlon) and nitrile butadiene rubber. See also ***poly-acrylonitrile*** and ***carbon fibre***.

Adrenaline is made in the body from either of two amino acids, phenylalanine or tyrosine, both of which have a benzene ring as part of their structure. Adrenaline can exist in two mirror image forms of which only the left-hand form is the natural and active one.

Amide bonds have the arrangement –NH–CO– and occur naturally in biopolymers, such as proteins, wool and silk, and in man-made polyamide polymers, such as nylon.

Amino acids have the general formula $RCH(NH_2)CO_2H$ and they join together to form polypeptides (proteins). There are 22 naturally occurring amino acids, eight of which are an essential part of the diet because they cannot be synthesised in the body. These are isoleucine, leucine, lysine, methionine, phenylalanine, threonine, tryptophan and valine.

Ammonium nitrate has the chemical formula NH_4NO_3 and is a colourless crystalline salt which is very soluble in water.

A Healthy, Wealthy, Sustainable World
By John Emsley

Published by the Royal Society of Chemistry, www.rsc.org

Anabolic steroid anabolic refers to metabolism and a steroid refers to a class of molecules with 17 carbon atoms arranged as 4 fused rings.

Anaerobic microbes thrive in a non-oxidising environment, in other words, an environment in which they are not in contact with the oxygen of the atmosphere.

Antibiotic is an antibacterial chemical produced by a living organism, usually by fungi which make them to protect themselves against bacteria. Chemists have been able to modify these natural molecules to make them suitable for humans to take, and have even been able to make new kinds of antibiotics.

Antioxidants include vitamins A, C and E, plus other substances, such as carotenoids and flavinoids. They are given this name because they react and neutralises potentially damaging ***free radicals*** which form when oxygen, O_2, and other oxygen-containing entities, lose an electron.

Benzoic acid has the formula $C_6H_5CO_2H$ and consists of an acid group (CO_2H) attached to a benzene ring [C_6H_5]. See also ***parabenzoic acid***.

Bandgap is the energy required to move an electron from a chemical bond to the conductance band, which is the region of space where it can move freely and thereby carry current. How much energy is required to do this determines whether the material is a conductor, semiconductor or an insulator. If the bandgap is less than 0.1 eV then the substance is a conductor, if it is between 0.1 and 3 eV it is a semi-conductor, and if it is above 3 eV it is an insulator. (eV is a tiny unit of energy called the electron volt and is the energy of an electron moving in an electric field of 1 volt.)

Barbituric acid (aka pyrimidine-2,4,6-trione; $C_4H_4N_3O_3$) is a cyclic molecule and was first prepared form urea and malonic acid by the German chemist Adolph von Bayer in 1863.

Barrel A barrel of oil contains 42 US gallons which is equivalent to 159 litres.

Biomass consists of the ***carbohydrates cellulose***, ***hemicellulose***, and ***lignin***. Biomass is also referred to as lignocellulose.

Bisphenol A (aka BPA) is made by reacting phenol with acetone and has the chemical formula HOC_6H_4–$C(CH_3)_2$–C_6H_4OH. It is used to manufacture ***polycarbonate***. It has been controversially linked to hormone imbalance and its use is restricted in some countries.

Blue-baby syndrome is more correctly called *methaemoglobinaemia* and refers to a medical condition in which the blood does not have enough oxygen because its red blood cells are contaminated with other molecules. Babies may be born in this condition. In blue-baby syndrome the blood does not have its normal rich red colour and the skin takes on a bluish pallor.

BMI is short for **body mass index** and it is calculated by dividing a person's weight in kg by their height in metres squared. The same formula applies to both men and women. A BMI below 20 means a person is underweight, between 20 and 25 is the ideal range, over 25 a person is referred to as overweight, and in excess of 30 is classed as obese. BMI is regarded as a guide to general health and an indicator of future health prospects.

Bordeaux mixture is made by mixing a solution of copper sulfate ($CuSO_4$) with one of hydrated lime ($Ca(OH)_2$) to give a final composition of one part by weight of the former to one part of the latter to 100 parts of water.

Butanol can exist in five different conformations: often referred to as *n*-butanol, *sec*-butanol (which can exist in two mirror image forms), *iso*-butanol and *tert*-butanol. All have the chemical formula C_4H_9OH but with different molecular structures. Normal or *n*-butanol has the chemical formula $CH_3CH_2CH_2CH_2OH$, secondary butanol or *sec*-butanol has the chemical formula $CH_3CH(OH)CH_2CH_3$ *iso*-butanol has the chemical formula $(CH_3)_2CHCH_2OH$, and tertiary butanol or *tert*-butanol has the chemical formula $(CH_3)_3COH$.

Caffeine is known chemically as 1,3,7-trimethylxanthine, meaning it has three methyl groups (CH_3) attached to a xanthine entity. Molecules with one less methyl group are theophylline, paraxanthine and theobromine, and behave similarly to caffeine. Theophylline is a slightly more powerful stimulant while theobromine is less so. Theophylline is prescribed for bronchial asthma in adult doses of 250 mg to relax the bronchia and ease breathing.

Carbohydrates are composed only of carbon, hydrogen and oxygen and they constitute a class of chemical compounds which includes simple sugars, such as glucose ($C_6H_{12}O_6$). The complexity of carbohydrates comes from the many ways in which smaller units like this can link together.

Carbon fibre consists of fibres that are 5–10 microns in diameter and these can be twisted together to produce a yarn that is incredibly strong and less heavy than steel wire. Most carbon fibre is manufactured by heating the polymer ***polyacrylonitrile*** (PAN) first at 300 °C and then at 2000 °C so it ends up completely carbonised. Carbon fibre yarn can even be woven to produce a fabric.

Carboxylic acids have the general formula RCO_2H in which H is the acidic hydrogen attached to an oxygen atom and R represents other atoms. The simplest carboxylic acid is formic acid HCO_2H (aka methanoic acid) where R is hydrogen, the next is acetic acid, CH_3CO_2H (aka ethanoic acid) and is the acid of vinegar. As the number of carbons of R increases so does the pungency of the acid, becoming particularly repulsive when R is a five or six carbon atom chain as in $C_4H_9CO_2H$, which is pentanoic acid (aka valeric acid), and which smells of faeces, and $C_5H_{11}CO_2H$, which is hexanoic acid (aka caproic acid), and which smells of goats.

Carboxymethylcellulose is cellulose in which most of its hydroxy groups (OH) have been replaced by OCH_2CO_2H groups. The acid hydrogen of the CO_2H can then be replaced by sodium, and such compounds make ideal thickening agents for various products including foods, and are coded E466. Carboxymethylcellulose as its sodium derivative is also used as an anti-redeposition agent in detergents, which means it binds to the fibres of cottons and linens, making them able to repel dirt that has been lifted off other items in the wash.

Cellophane is ***cellulose*** **acetate** and is a transparent cellulose polymer which has been widely used as packaging film and for wastewater treatment membranes.

Cellulose consists of long chains of glucose rings joined together in straight rows. Cellulose acetate can be made from this by replacing the hydrogen atoms of the OH groups along the chain with acetyl groups ($OCCH_3$) by reaction with acetic anhydride. Cellulose acetate is made from the cellulose of wood pulp and is used for fabrics, plastics and the filters of cigarettes.

Chirality is the feature exhibited by a molecule when it can exist as two distinct mirror image structures. Just as we have left and right hands, so there can be left and right molecules.

Chlorodifluoromethane ($CHClF_2$) is a gas that was previously used as a replacement for CFCs which are damaging to the ozone layer. However, it is a potent greenhouse gas and on that account it is now being phased out where its use would result in its release to the environment.

Chromium as chromium (III) is an essential trace element for the human body, whereas in its higher oxidation state as chromium (VI) it is dangerous and is a known carcinogen.

Coagulants such as aluminium sulfate and iron(III) sulfate are added to dirty water and the insoluble hydroxides these form, namely $Al(OH)_3$ and $Fe(OH)_3$, absorb impurities as they settle out. At one time it was thought that they worked by virtue of their highly charged cations, namely Al^{3+} and Fe^{3+}, but it now appears that polymeric cations such as $[Al_{13}O_4(OH)_{24}]^{7+}$ and $[Fe_3(OH)_4]^{5+}$ are the active species.

Deuterium chemical symbol D, is the heavier form of hydrogen. A hydrogen atom has a singe positive proton in its nucleus whereas deuterium has both a proton and a neutral neutron. The neutron is as heavy as the proton and this doubles the atom's weight and slightly changes the properties of the compound in which D replaces H, none more so than in D_2O, aka heavy water.

Dithane is an insecticide based on dithiocarbamic acid (H_2NCS_2H) as either its zinc or zinc manganese salts, and it is sold under various trade names.

Dichloroisocyanurate (sodium) has the chemical composition $C_3Cl_2N_3O_3Na$ and consists of a ring of three alternate carbon and nitrogen atoms with the oxygen atoms attached to the carbons and the two chlorines attached to two of the nitrogens.

Double-blind test is one in which neither the person receiving a drug nor the person giving it to them knows whether it is the active ingredient or a placebo. Double-blind tests ensure that the person giving the drug cannot reveal its composition to the recipient.

Electrospray ionisation mass spectrometry is a way of delivering very large molecules to a mass spectrometry in an aerosol format which keeps them intact. Earlier methods for introducing such molecules into a mass spectrometer invariably resulted in their breaking up into fragments.

Epichlorohydrin (aka chloromethyloxirane) has the chemical formula C_3H_5ClO and consists of a three-membered ring of one oxygen and two carbons with a CH_2Cl group attached to one of the carbons. It is a liquid and highly reactive and used in the production of plastics and elastomers.

Epidemiology is a method by which the statistical analysis of a large mass of data collected from a large population will reveal links that are not apparent when only small numbers are considered. The numbers of persons involved should be large (ideally in their thousands) and they should be compared with a correspondingly large group of those without the condition under investigation. The groups should be balanced in terms of gender, ages, average body weight, ethnicity, personal habits (*e.g.* smoking, alcohol consumption), social class, occupation, location, *etc.* Unless great care is taken, epidemiology can reveal spurious links because of confounding variables which have been overlooked, and it is this which makes the majority of such findings of little value except for publicity purposes.

Fatty acids in most oils and fats are palmitic acid ($C_{15}H_{31}CO_2H$), stearic acid ($C_{17}H_{35}CO_2H$), oleic acid ($C_{17}H_{33}CO_2H$) which has one double bond along its chain, and linoleic acid ($C_{17}H_{31}CO_2H$) with two double bonds along its chain. Some plant seeds, notably those of rape, flax (whose oil is known as linseed oil) and soya, also produce linolenic acid ($C_{17}H_{29}CO_2H$) which has three double bond along the hydrocarbon chain.

Fatty alcohol ethoxylates formula $CH_3(CH_2)_{9-11}OCH_2CH_2OCH_2CH_3$, are used a mild, non-ionic, surfactants.

Fischer–Tropsch is a process for producing hydrocarbons from ***syngas*** and the overall reaction is summarised by the equation:

$$(2n + 1)H_2 + nCO \rightarrow C_nH_{2n}+2 + nH_2O.$$

Fossil fuels are natural gas, oil and coal, and they are the carbon-based remains of biomass growing in the Carboniferous era hundreds of millions of years ago. Natural gas is mainly methane (CH_4), oil consists of many different molecules with approximate composition $(CH_2)_n$, while coal consists of a complex array of interlinked molecules with approximate composition $(CH)_n$.

Free radicals are atoms or molecules with at least one unpaired electron and these are particularly active chemical entities because they seek to pair this electron by abstracting one from another molecule. Oxygen (O_2) has two unpaired electrons.

Fuel cells are electrochemical devices for generating useful electricity and they run on hydrogen gas. The hydrogen molecule (H_2) enters the anode compartment where a platinum catalyst breaks it into hydrogen atoms and these are separated from their electrons, which move off to provide the current to drive an electric motor. The hydrogen nuclei, which are simply protons, move through an electrolyte and a membrane towards a cathode where they meet up with oxygen to form water.

Gallium nitride has the formula GaN and it has some remarkable properties as a semiconductor material in that, unlike silicon, it does not have to be cast in the form of absolutely perfect crystals in order for it to work. Blu-ray technology has been built on this remarkable material.

Gasoil is the mix of hydrocarbons which distils from crude oil when it is heated between 200–300 °C (*ca.* 400–600 °F).

GC-MS stands for gas chromatography-mass spectrometry. It is one of the most sensitive forms of analysis and is capable of identifying minute traces of chemicals in, for example, a single strand of hair. The chromatography carefully separates all the different component molecules in a mixture and the mass spectrometer identifies each of them.

Glucose is the simple carbohydrate, $C_6H_6(OH)_6$, with a six membered ring.

Glycerol (aka glycerine and propane-1,2,3-triol) is a by-product of biodiesel manufacture and for every litre of biodiesel produced there is 90 g of glycerol, totalling more than 1 million tonnes globally per year. Some is added to cough syrups, lotions, skin care products, shaving cream, and hair gels. It can also be used to make the explosive nitroglycerine. Glycerol can be a raw material for the chemical industry and there are several ways of converting it to more useful raw materials such as ***epichlorohydrin***. The Belgian chemical company Solvay has an award-winning process for doing this.

Graphene is a type of carbon in the form of single sheets of carbon atoms in a honeycomb arrangement of interconnected benzene rings.

Hemicellulose consists of many difference kinds of carbohydrates, and in particular ones with five carbon atoms such as galactose, mannose, xylose and arabinose. Unlike *cellulose* whose long chains consist of only glucose (which has six carbon atoms) hemicellulose chains are much shorter and with many side chains. It easily broken up by reaction with acids.

HFCs is short for hydrofluorocarbons and they consist only of carbon, hydrogen and fluorine atoms and therefore do not damage the ozone layer as do ***CFCs***. Nevertheless, they are powerful greenhouse gases.

Hydrocarbons are chemicals made entirely of carbon and hydrogen. Those with one to four carbons are gases. The hydrocarbons with around seven or eight carbon atoms are the basis of petrol (aka gasoline), while those with around ten are used for diesel, and those with 12 to 16 carbons, are the paraffin (aka kerosene) that fuels most aircraft. As the chains get even longer the liquids become viscous fluids, ideal for including in cosmetics, then eventually they are greases, like Vaseline, and finally they are solid waxes, suitable for making smokeless candles.

Hydrogen peroxide has the chemical formula H_2O_2 and is a powerful oxidising agent.

Hypochlorite *see* ***sodium hypochlorite***.

Immunoassay is a method of analysis using antibodies and enzymes to identify a drug and this method can work for certain substances which are not easily tested for by other methods.

Indium tin oxide (aka ITO, In_2SnO_5) is transparent, colourless and conducts electricity. Deposited as a thin layer on the surface of glass in solar cells it can act as an electrode while not impeding the passage of the sunlight which generates the electricity.

In vitro* and *in vivo are terms used respectively to describe experiments which are done in laboratory apparatus and on living things. The terms are derived from the Latin words meaning *in glass* and *in life*.

Ion exchange resins are tiny beads around 1 mm in size made from polymers. The beads have a charged surface and can exchange one ion for another, such as softening water by replacing calcium with sodium, or they can be used to remove all dissolved metal and non-metal ions from water by replacing them with the H^+ or OH^- which together form water. Periodically an ion exchanger has to be regenerated by reverse ionic exchange, replacing metals with H^+ and non-metal anions with OH^-. (There are now ion exchange resins which can be recharged electronically and much more quickly.)

Invert sugar is made from ordinary sugar which is sucrose. Sucrose consists of a ***glucose*** molecule attached to a fructose molecule. Under the action of the enzyme *invertase,* or of an acid, sucrose splits into its two components, of which fructose is intensely sweet. It is then known as invert sugar.

Isoprene (aka 2-methyl-1,3-butadiene) is a colourless liquid formula $CH_2{=}C(CH_3)CH{=}CH_2$.

Kevlar (aka polyparaphenylene terephthalamide and PPTA) consists of long strands of benzene rings interconnected with amide (–NH–CO–) groups. This polymer's remarkable strength comes from its regular structure with its stands lining up into parallel rows and flat sheets.

Lactic acid (aka 2-hydroxypropanoic acid) gets its original name from *lac* the Latin word for milk, and it used to be known as milk acid. Its chemical formula is $CH_3CH(OH)CO_2H$. This simple molecule exhibits ***chirality***. Its two forms are known as *D*-lactic acid and *L*-lactic acid (where *D* stands for *dexter* from the Latin for right, and *L* stands for *laevus*, meaning left). Industrially produced lactic acid is normally a mixture of the *D* and *L* forms, but in living things one form tends to dominate, and the human body produces mainly *L*-lactic acid, which is present in blood, muscle, liver, kidneys and other organs.

Lanthanide metals (aka lanthanoids or rare earth metals) range from element 57 (lanthanum, La) to element 71 (lutetium, Lu). Some of these are not particularly rare, such as cerium (element 58, Ce), while some really are very rare, such as thulium (element 69, Tm), and one of the lanthanides is no longer present on Earth. Promethium is radioactive is all isotopes of which the longest lived, promethium-145, has a half-life of only 17.7 years. Notwithstanding their relative rarity, uses are now being found for lanthanide elements such as in low energy light bulbs. Chemically the lanthanides are very similar to one another.

LC-MS is liquid chromatography linked to mass spectrometry and like **GC-MS** it is capable of extremely sensitive chemical analysis.

Light-emitting diode (LED) emits light when the current flows and this happens when (negative) electrons flowing in one direction meet positive 'holes' moving towards them. As they combine they emit a burst of light energy and its wavelength and colour depends on the semiconductor from which the diode is made.

Lignin is a major component of wood and consists of benzene rings interlinked through short carbon chains. The benzene rings also have other groups like methoxy (OCH_3) attached to them. Lignin is an important component of wood (around 30%) and is part of its cell walls, as it is of those of other plants. Lignin has to be removed from wood pulp before it can be turned into paper.

Lipopeptides are cyclical peptide molecules (see ***amino acids***) with a long hydrocarbon chain attached.

Liquid crystal molecules all must have a *polar* head and a *linear* tail. The polar head means that it has a positive part and a negative part rather like poles of a magnet. There are difference types of liquid crystals. Those which make liquid crystal displays are those with a tendency to line up in parallel. There are also twisted ones which unwind when exposed to a voltage, the degree of unwinding being directly related to the voltage they experience. Some liquid crystals respond to changes in temperature – and are used as thermometer strips – and these are called thermotropic.

Lithium batteries can produce twice the voltage of ordinary zinc–carbon batteries. There are various kinds but the most popular has lithium iodide as the anode (+ end) and manganese dioxide as the cathode (−) with lithium perchlorate in propylene carbonate and dimethoxyethane as the electrolyte. As such they deliver 3 volts.

Lithium–iodine batteries are more reliable and they have lithium as the anode and iodine as the cathode with poly(2-vinylpyridine) (aka P2VP) between them. These batteries supply a low current at 2.8 volts and last for years. Some people have had them implanted in pacemakers for as long as 15 years.

Malic acid (chemical formula $C_4H_6O_5$, aka E296) has two carboxylic acid groups and is responsible for the sourness of unripe fruits and especially apples. It disappears as fruit ripens, being converted to the milder lactic acid. Malic acid can exist as both left (L) and right (D) molecules although natural sources produce only L-malic acid.

Mannitol is a carbohydrate of formula $C_6H_6(OH)_6$ which is used medically as a diuretic.

Megapixel is a million pixels. A screen needs to have as many pixels as possible. A small screen will have a grid of pixels 1024 by 768, which is 786 432 pixels in total. These are not arbitrary numbers but are powers of two, namely 2^{10} and 1.5×2^9. Increase these by doubling them to 2^{11} by 1.5×2^{10} and there are now 3.15 *mega*-pixels – or 3 145 728 to be exact.

Melamine formula N_3C_3 $(NH_2)_3$, is an industrial chemical, which is used to make plastics, glues and flame retardants.

Melatonin (chemical name: N-acetyl 5-methoxy-tryptamine) is pale yellow, with leaf-like crystals which melt at 117 °C. The pineal gland produces it from serotonin, the brain chemical that regulates mood, and this in turn is made from the essential ***amino acid*** tryptophan.

Microgram is a millionth of a gram and about the weight of a spec of dust.

Naphtha is the mix of hydrocarbons which distils from crude oil when it is heated between 65–200 °C (*ca.* 150–400 °F).

Nitrile rubber is a copolymer of acrylonitrile and butadiene and it is the material of which blue and purple disposable gloves are made. (Latex gloves are white.) Nitrile rubber has the advantage of not being affected by oil and chemical solvents so it is widely used in the automotive industries for hoses, seals and grommets.

N*-methyl morpholine oxide** has the chemical formula $C_5H_{11}NO_2$ in which the nitrogen atom has attached to it a methyl group (CH_3) and an oxygen atom. Although it is a solid, when it is molten it has the ability to dissolve ***cellulose.

Noble gas is the name given to the members of group 18 of the periodic table and they are called noble because they are chemically very unreactive. They are helium, neon, argon, krypton, xenon and radon. Argon is the most abundant and comprises 1% of the atmosphere. The others are present in only tiny amounts.

Nylon is a polyamide polymer, in other words, its chains contain the amide link which is –C(O)–NH–. Nylon fibres are used in fabrics and rope, while solid nylon is used for plastic screws, gears, and other mechanical items which are not exposed to high stress levels.

Oligosaccharides are carbohydrates which are supposed to nourish 'good' bacteria in the human gut. There are three kinds: fructooligosaccharides (FOS), galactooligosaccharides (GOS), and lactulose. FOS consists of a chain of fructose units, as many as 60 in some cases. GOS consists of two linked galactose units plus a glucose unit. Lactulose consists of just a galactose and fructose joined together.

Omega-3 is used to describe *fatty acids* in which the double bond in the hydrocarbon chain is between the third and fourth carbons from the free end of the chain. **Omega-6** means the double bond is between the sixth and seventh carbons.

Organic light-emitting diodes (OLEDs) consist of a cathode (negative electrode) and an anode (positive electrode) between which is a double layer of polymer, one layer of which accepts electrons from the cathode while the other layer next to the anode loses electrons to it thereby creating positive 'holes'. The electrons in the cathode layer attract the 'holes' in the anode layer and as they combine they emit light.

Orlistat is a natural inhibitor of pancreatic lipase enzymes and it is isolated from the bacteria *Streptomyces toxytricini*. It is sold as alli and Xenical slimming aids. It is derived from lipstatin, a natural product that was first extracted from the bacterium *Streptomyces toxytricini* in 1983. Lipstatin is not very stable but can be made so by reacting its two double bonds with hydrogen and that is what orlistat is.

Ozone has the formula O_3 and consists of three oxygen atoms joined together. This gas is part of the atmosphere, but in very low concentrations. High in the atmosphere it is referred to as the ozone layer, protecting the planet against damaging UV rays from the sun. At ground level it is more of an atmospheric pollutant. Ozone is used to disinfect water as an alternative to ***hypochlorite***.

Parabenzoic acid (aka *para*-hydroxybenzoic acid or 4-hydroxybenzoic acid) has the chemical formula $HOC_6H_4CO_2H$, in which the OH group is at the opposite end of the benzene ring [C_6H_4] to the CO_2H group.

Paraquat is the fast-acting herbicide *N,N′*-dimethyl-4,4′-bipyridinium dichloride. It kills all green-leaf plants on contact and is widely used throughout the world. It is toxic to animals and contains dyes and bitter-tasting additives to prevent misuse.

PFTE stands for polytetrafluoroethylene. It has the chemical formula –$(CF_2CF_2)_n$–. Its popular name is Teflon.

Photovoltaic (PV) power devices use semiconductor silicon in two layers, an electron-rich layer on top of an electron-poor layer. These produce an electric field at the junction between them. The electron-rich layer is silicon doped with a few atoms of phosphorus to give it a surplus of electrons and is known as the n-layer (negative layer), while the electron-deficient layer is silicon doped with a few atoms of boron to give it a deficit of electrons and this is known as the p-layer (positive layer). The n-layer releases electrons when exposed to light; while the p-layer provides 'holes', which are the gaps in chemical bonding where electrons are missing. Together these create a voltage difference which causes current to flow.

Polyamide is a polymer joined through amide bonds (–C(O)–NH–) and they can either be biopolymers, such as wool and silk, or man-made polymers, such as nylon. Polyamide membranes are strong and are used in wastewater treatment and reverse osmosis desalination plants.

Polycarbonate is manufactured from ***bisphenol A*** and carbonyl chloride ($COCl_2$). It has the chemical structure $(–C_6H_4–C(CH_3)_2–C_6H_4–O–CO–O–)_n$ [C_6H_4 is a benzene ring].

Polyetheretherketone (aka PEEK and Victrex) is made by reacting 4,4′-difluorobenzophenone with the sodium salt of hydroquinone at 300 °C. It has the chemical structure $(–O–C_6H_4–CO–C_6H_4–O–C_6H_4–)_n$.[C_6H_4 is a benzene ring].

Poly(hydroxybutyrate) (aka PHB) is a polymer produced by bacteria and extractable from them. Its structure is $-(CH(CH_3)-CH_2-CO)_n-$. The methyl group on the carbon can also be an ethyl group in which case the polymer is known as poly(hydroxyvalerate) or PHV. These kinds of polymers are referred to collectively as poly(hydroxyalkanoate) or PHA.

Polymers are molecules which repeat the same unit in a seemingly endless chain. The simplest polymer is polyethylene, in which ethylene molecules are linked to form $–CH_2–CH_2–CH_2–CH_2–$ chains. Other polymers have different atoms and groups attached to the carbons, and some differ in having oxygen, nitrogen or sulfur atoms as part of the chain.

Polymers are the basis of many plastics and fibres. The term 'poly' means many and refers to the fact that they consist of repeating units of the same molecule joined together in long chains. If the polymer chains are joined at intervals along their length to neighbouring chains they are called cross-linked, and this makes them rigid. Some polymers are derived from mixtures of starting materials and are said to be copolymers.

Polymers that are stretchable are called elastomers.

Polymers which are plastics are either thermosets or thermoplastics. Thermosets are chemically altered by heat because this leads to cross-linking. The resulting products set hard and remain so. Thermoplastics are not permanently affected by heat; they soften when heated and harden on cooling.

The common polymers, codes and common uses are given in the following table:

Polymer	*Code*	*Product applications*
Polyethylene	PE	see below
Low-density polyethylene	LDPE	plastic film, bags, coating for paper
High-density polyethylene	HDPE	moulded products, containers, crates
Polypropylene	PP	film, carpets, clothing, bottles
Polystyrene	PS	packaging, toys, cups, cutlery
Expandable polystyrene	EPS	insulation, packaging, aggregates
High-impact polystyrene	HIPS	picnic ware, food tubs
Polyethylene terephthalate	PET	bottles, food trays, duvet filling
Polyvinyl chloride	PVC	window frames, flooring, tubing, pipes
Styrene butadiene rubber	SBR	tyres, footwear, moulded goods
Polyurethane	PU	foam padding, surfaces, elastomers

Polymethylmethacrylate is known more correctly as poly(methyl 2-methylpropenoate). It is a polymer noted for its strength and transparency and it goes under the name Perspex in the UK and Lucite in the USA.

Polypropylene is the polymer of propene, which has the formula ($CH_3CH{=}CH_2$), and which can be polymerised by heating it under a pressure of 15 atmospheres at temperatures from 50 to 90 °C. The result is a chain of carbon atoms with methyl groups attached to alternate ones. These methyl groups can be arranged in three different ways. All can be at the same side, or alternatively one side and then the other, or randomly positioned, and this offers variety in properties of the final product.

Polysaccharides To make rayon or cellophane, ***cellulose*** from wood pulp is dissolved in sodium hydroxide solution and carbon disulfide which produces a viscous solution which is then forced through a tiny hole, to give rayon, or through a slit to give cellophane. These are then passed through dilute sulfuric acid and various other washes to give the final material.

Polystyrene has the formula $-(CH(C_6H_5)-CH_2)_n-$. [C_6H_5 is a benzene ring.] It was discovered as long ago as 1839, but only went into commercial production in 1930. It is made from styrene monomer, an oily liquid which boils at 145 °C and it can be polymerised by heating a suspension of it in water. The product is small beads of polystyrene which are then graded into various sizes ranging from 0.2 to 3 mm, which can be softened with steam and shaped or moulded.

Polysulfone is a rigid, transparent, tough plastic formed from the reaction of ***bisphenol-A*** with bis(4-chlorophenyl)sulfone. Its chemical formula is $-(O-C_6H_4-C(CH_3)_2-C_6H_4-O-C_6H_4-SO_2-C_6H_4)_n-$ [C_6H_4 is a benzene ring]. It is unaffected by acids, alkalis and oxidising agents but it softens and slowly dissolves when exposed to some solvents. It is ideal for making membranes and is used in dialysis equipment, wastewater recovery and gas separation.

Polyunsaturated fatty acids are long chain fatty acids with two or more double bonds along the chain.

Polyurethane is formed from polyols and isocyanates. The commonly used isocyanates are either toluene diisocyanate (TDI) or diphenylmethane diisocyanate (MDI), but it is the variety of polyols that leads to such wide applications for this polymer. Low molecular weight polyols with lots of OH groups give products that have high abrasion resistance and find use in coating undersea pipelines and in security glazing. If a volatile liquid is also present when the polyurethane forms it will generate bubbles in the plastic, expanding it as it sets, like a sponge cake in an oven. Depending on the chemicals used, and the extent of bubbling, the final product can be a flexible foam, ideal for furniture, or a rigid foam, suitable for fridges and wall insulation.

Polyvinyl alcohol (aka PVOH) has the formula $-(CH(OH)CH_2)_n-$ and is the basis of water-soluble packaging.

ppm stand for parts per million and is equivalent to 1 milligram in 1 litre.

PVC (aka polyvinyl chloride) has the chemical formula $-(CH(Cl)CH_2)_n-$.

Rayon see ***polysaccharide***.

Rotenone (chemical formula $C_{23}H_{22}O_6$) is a complex molecule of five interlinked rings. It occurs naturally in the roots of various plants, some of which like the tuba plant can be dried and powdered to yield derris which is used as an insecticide. Retenone also kills fish although is only slightly toxic as far as mammals are concerned.

Sodium hypochlorite (aka bleach) is one of the most powerful disinfectants. Its chemical formula is NaOCl and it is produced by bubbling chlorine up a column, down which trickles a solution of sodium hydroxide. Hypochlorite is a strong oxidising agent and is stable. Because it is made from chlorine, ordinary bleach is sometimes wrongly called chlorine bleach, but it contains no free chlorine gas as such.

Starch is a component of seeds, tubers and roots. It has two major components: amylose and amylopectin. Amylose is a polymer consisting of linked ***glucose*** molecules and a typical strand contains around 2500 of them. Amylopectin has even longer chains with lots of shorter carbohydrate side chains, and this type of starch makes it a good thickening agent for foods like sauces.

Surfactants are molecules which have the ability to solubilise things like greases and oils in water. This they do by having a part of the molecule that is attracted to water and part that is attracted to the grease and the oil. There are four kinds of surfactant and they vary according to their water-seeking end which can be negatively charged (and known as anionic surfactants), positively charged (cationic), both negative and positive (amphoteric), or neutral (non-ionic). Soap is an anionic surfactant.

Syngas is a mixture of hydrogen and carbon monoxide. The reaction for syngas from natural (methane) gas is $CH_4 + H_2O \rightarrow 3H_2 + CO$, while that which forms syngas from coal, coke or charcoal is

$C + H_2O \rightarrow H_2 + CO$. Syngas can also be formed directly from biomass by reaction with steam.

TAED (aka tetraacetylethylenediamine) has the formula $(CH_3CO)_2NCH_2CH_2N(COCH_3)_2$. It acts as an activator of the peroxide bleaches used in detergents which it does this by reacting to form peracetic acid (CH_3CO_3H), a powerful oxidising agent.

Teflon is the common name for polytetrafluoroethylene (aka ***PTFE***), which is made by polymerising tetrafluoroethene, ($CF_2{=}CF_2$).

Titanium dioxide has the formula TiO_2 and is a brilliant white material used in paints and foods.

Trichloroisocyanuric acid has the chemical formula $C_3O_3N_3Cl_3$ and consists of a six-membered ring of alternate carbon and nitrogen with the oxygens on the carbons and the chlorines on the nitrogens. It is a white, crystalline powder that is used commercially as a disinfecting agent and bleaching agent because it forms ***hypochlorite*** in water.

Triclosan (chemical name 5-chloro-2(2,4-dichlorphenoxy)phenol, $C_{12}H_7Cl_3O_2$) is a powerful antibacterial and antifungal agent used in many personal hygiene products such as soaps, deodorants, and mouthwashes. It is powerful enough to kill the deadly MRSA bacteria. Some worry that its overuse might result in resistant strains of microbes, while others see it as an environmental threat.

Triglyceride is the more scientific name for fats and oils and these consist of three hydrocarbon fatty acids attached to a glycerol molecule. The fatty acids can be saturated, monounsaturated or polyunsaturated and in theory they could all be included in a single triglyceride molecule although naturally occurring triglycerides tend to be predominantly of one kind.

Units used in this book are a mixture of Imperial, US and SI. Most are based on the SI system of weights and measures and for weights they are tonnes (t), kilograms (kg), grams (g), milligrams (mg), micrograms (μg) and nanograms (ng). Each one of these is a thousand times smaller than the one preceding it. The units of measure are metres (m) and kilometres (km).

Where there are different systems, then a few simple conversions should be noted. Thus an Imperial pound is equivalent to 448 g (0.448 kg) and an Imperial ton is 1016 kg.

An imperial gallon is equivalent to 4.545 litres, while a US gallon is 3.785 litres.

An acre is a measure of land area and corresponds to 0.405 hectare. A hectare is 10 000 square metres.

Vinyl chloride (aka chloroethene) has the chemical formula $CH_2{=}CHCI$ and is a gas.

Visible light has wavelengths in the range 380 nm (violet) to 750 nm (red). The colours of the spectrum are red (620–750 nm), orange (590–620 nm), yellow (570–590 nm), green (495–570 nm), blue (450–495 nm) and violet (380–450 nm). The supposed seventh colour, indigo, is no longer regarded as a distinct colour lying between blue and violet.

Zinc dialkyldithiophosphates have the general formula $Zn[S_2P(OR)_2]_2$ where R is an linear hydrocarbon chain.

Sources and Further Reading

Much of the information in *A Healthy, Wealthy, Sustainable World* is derived from articles published within the past five years in *Chemical & Engineering World* (the weekly magazine of the American Chemical Society), *Chemistry & Industry* (the fortnightly magazine of the Society of Chemical Industry) and *Chemistry World* (the monthly magazine of the Royal Society of Chemistry). The Royal Society of Chemistry also publishes reports which address the issues discussed in this book and some of these are listed below. For readers who are not chemists but would like to know more about chemistry's role in everyday life, there is a list of popular science books on the subject. Finally there are books which I would recommend for those who, like me, are worried that as a society we seem to be moving in the wrong direction, and one in which science is either ignored, regarded with suspicion, or wilfully misunderstood.

Data Books

While a great deal of information can be obtained from the internet, books are more reliable because they have been checked by several people and have been subject to review. (The same is true of websites of official organisations.) Here are some books which I found useful.

H.-D. Belitz and W. Grosch, *Food Chemistry*, Springer Verlag, 1986.

W. Büchner, R. Schliebs, G. Winter and K. H. Büchel, *Industrial Inorganic Chemistry*, VCH, Weinheim, Germany, 1989.

S. Budavari (ed.), *The Merck Index*, Merck & Co. Inc., Rahway NJ, USA, 11th edn, 1989.

A Healthy, Wealthy, Sustainable World
By John Emsley

Published by the Royal Society of Chemistry, www.rsc.org

F. Cardarelli, *Materials Handbook*, Springer Verlag, London, 2000.
T. P. Coultate, *Food: The Chemistry of its Components*, Royal Society of Chemistry, Cambridge, 5th edn, 2008.
K. Diem and C. Lentner (ed.), *Scientific Tables*, Documenta Giegy, Basle, Switzerland, 7th edn, 1970.
Dietary Reference Values for Food Energy and Nutrients for the UK, Department of Health, HMSO, London, 1991.
J. Henry (ed.), *BMA Guide to Medicines and Drugs*, Dorling Kindersley, London,1993.
R. J. Kutsky, *Handbook of Vitamins, Minerals and Hormones*, Van Nostrand Reinhold, New York, 2nd edn, 1981.
A. Paul and D. A. T. Southgate, *McCance and Widdowson's The Composition of Foods*, Her Majesty's Stationery Office, London, 1978.
D. A. J. Rand and R. M. Dell, *Hydrogen Energy: Challenges and Prospects*, RSC Publishing, Cambridge, 2008.
T. Stone and G. Darlington, *Pills, Potions, and Poisons: How Drugs Work*, Oxford University Press, Oxford, 2000.
V. Smil, *Enriching the Earth: Fritz Haber, Carl Bosch and the Transformation of World Food*, MIT Press, Boston, Mass, 2001.
Wade (ed.), *Martindale, the Extra Pharmacopoeia*, The Pharmaceutical Press, London, 1977.
D. H. Watson (ed.), *Natural Toxicants in Food*, Ellis Horwood/VCH, 1987.

The Royal Society of Chemistry Reports and Books

The RSC regularly publishes papers on a diverse range of topics and these can be accessed *via* its website http://www.rsc.org/ScienceAndTechnology/Policy/. The following ones relate to topics covered in this book and which have been published in recent years.

Reducing Carbon Emissions from Transport, February 2006
Investigating the Oceans, February 2007
Sustainable Water: Chemical Science Priorities, November 2007
Fuelling the Future, November 2007
Renewable Electricity Generation Technologies, January 2008
Harnessing Light: Solar Energy for a Low Carbon Future, January 2008
Proposal for a European Regulation on Novel Foods, June 2008
Why Do We Worry About Phthalates? December 2008
Green Chemistry, December 2008
The Vital Ingredient: Chemical Science and Engineering for Sustainable Food, January 2009.

UK Expertise for Exploitation of Biomass-Based Platform Chemicals, March 2009
Why do we worry about Chemicals? April 2009
Integrated Pollution Prevention and Control (IPPC), May 2009
D. A. J. Rand and R. M. Dell, *Hydrogen Energy: Challenges and Prospects*, RSC Publishing, Cambridge, 2008.

Popular Science Books about Chemistry

There are scores of popular science books and some are about chemistry.

P. Atkins, *Molecules*, Cambridge University Press, Cambridge, 2003.
P. Atkins, *The Periodic Kingdom*, Weidenfeld & Nicolson, London, 1995.
P. Ball, *Stories of the Invisible*, Oxford University Press, Oxford, 2001.
P. Ball, *The Ingredients*, Oxford University Press, Oxford, 2002.
P. Ball, *Elegant Solutions*, Royal Society of Chemistry, Cambridge, 2002.
P. Ball, *The Elements: A Very Short Introduction*, Oxford University Press, Oxford, 2004.
P. Ball, *Elegant Solutions*, Royal Society of Chemistry, Cambridge, 2002.
P. Le Couteur and J. Burreson, *Napoleon's Buttons*, Penguin Putnam, New York, 2003.
J. Emsley, *Consumer's Good Chemical Guide*, Oxford University Press, Oxford, 1994.
J. Emsley, *Molecules at an Exhibition*, Oxford University Press, Oxford, 1998.
J. Emsley, *Shocking History of Phosphorus*, Macmillan, London, 2000.
J. Emsley, *Nature's Building Blocks*, Oxford University Press, Oxford, 2001.
J. Emsley, *Vanity, Vitality, Virility*, Oxford University Press, Oxford, 2004.
J. Emsley, *Better Looking, Better Living, Better Loving*, Wiley-VCH, Weinheim, 2007.
J. Emsley, *Molecules of Murder*, Royal Society of Chemistry, Cambridge, 2008.
S. Garfield, *Mauve*, Faber and Faber, London, 2000.
H. B. Gray, J. D. Simon and W. C. Trogler, *Braving the Elements*, University Science Books, Sausalito, California, 1995.

K. Karukstis and G. R. Van Hecke, *Chemistry Connections*, Academic Press, San Diego, 2000.
P. Marshall, *The Philosopher's Stone*, Macmillan, London, 2001.
J. Schwarcz, *Let Them Eat Flax*, ECW Press, Toronto, 2005
J. Schwarcz, *The Genie in the Bottle*, ECW Press, Toronto, 2005
J. Schwarcz, *Brain Fuel: 199 Mind-Expanding Inquiries into the Science of Everyday Life*, Doubleday, Canada, 2008.
J. Schwarcz, *An Apple A Day: The Myths, Misconceptions, and Truths About the Foods We Eat*, Other Press, New York, 2009.
B. Selinger, *Chemistry in the Market Place*, Harcourt Brace Jovanovich, Sydney, Australia, 5th edn, 1998.
B. Selinger, *Why the Watermelon Won't Ripen in Your Armpit*, Allen & Unwin, St Leonards, NSW, Australia, 2000.
C. H. Snyder, *The Extraordinary Chemistry of Ordinary Things*, John Wiley & Sons, Inc., New York, 1992.
P. Strathern, *Mendeleyev's Dream: the Quest for the Elements*, Hamish Hamilton, London, 2000.

Books which Support Science and Question Non-Science Attitudes in the Modern World

This century will have to see more support for science if we are to solve the world's most pressing problem of protecting the planet, and that means changing peoples' ideas and attitudes. At this stage, the emphasis is on showing that current beliefs in alternative viewpoints are leading us in the wrong direction. The following books are to be recommended for their exposure of the delusions that have led many well-intentioned people astray.

M. Allaby, *Facing the Future: The Case for Science*, Bloomsbury, London, 1995.
R. Bate (ed.), *What Risk? Science, Politics and Public Health*, Butterworth Heinemann, Oxford, 1997.
C. Booker and R. North, *Scared to Death*, Continuum, London, 2007.
I. Carson and V. V. Vaitheeswaran, *Zoom: The Global Race to Fuel the Car of the Future*, Penguin Books, London, 2008.
S. Feldman and V. Marks (ed.), *Panic Nation*, John Blake, London, 2005.
B. Goldacre, *Bad Science*, Fourth Estate, London, 2008.
W. Gratzer, *The Undergrowth of Science: Delusion, Self-Deception and Human Frailty*, Oxford University Press, Oxford, 2000.

M. Henderson, *Living with Risk*, The British Medical Association and John Wiley & Sons, Chichester, UK, 1987.

D. J. C. MacKay, *Sustainable Energy – Without the Hot Air*, UIT, Cambridge, 2009.

S. J. Milloy, *Junk Science Judo*, Cato Institute, Washington DC, 2001.

T. Murcott, *The Whole Story: Alternative Medicine on Trial*, Macmillan, London, 2005.

R. Park, *Voodoo Science: The Road from Foolishness to Fraud*, Oxford University Press, Oxford, 2001.

Sokal and J. Bricmont, *Intellectual Impostures*, Profile Books, London, 1998.

D. Taverne, *The March of Unreason*, Oxford University Press, Oxford, 2005.

V. V. Vaitheeswaran, *Power to the People*, Earthscan, London, 2005.

F. Wheen, *How Mumbo-Jumbo Conquered the World*, Fourth Estate, London, 2004.

Subject Index